PENGUIN BOOKS

HERE ON EARTH

Ti͏ ͏lannery is an internationally acclaimed scientist, explorer and conserva-
t͏ ͏ ͏ described by Sir David Attenborough as being 'in the league of the
͏ ͏great explorers like Dr David Livingstone'. His books include *The
 Eaters*, *The Eternal Frontier*, *Throwim Way Leg*, *A Gap in Nature, Country
 ͏e Weather Makers* ('one of the most influential books of the twentieth
 y' *Guardian*)

HERE
ON
EARTH

A TWIN BIOGRAPHY
OF THE PLANET AND THE
HUMAN RACE

TIM
FLANNERY

PENGUIN BOOKS

PENGUIN BOOKS

Published by the Penguin Group
Penguin Books Ltd, 80 Strand, London WC2R 0RL, England
Penguin Group (USA) Inc., 375 Hudson Street, New York, New York 10014, USA
Penguin Group (Canada), 90 Eglinton Avenue East, Suite 700, Toronto, Ontario,
Canada M4P 2Y3 (a division of Pearson Penguin Canada Inc.)
Penguin Ireland, 25 St Stephen's Green, Dublin 2, Ireland (a division of Penguin Books Ltd)
Penguin Group (Australia), 250 Camberwell Road, Camberwell, Victoria 3124, Australia
(a division of Pearson Australia Group Pty Ltd)
Penguin Books India Pvt Ltd, 11 Community Centre,
Panchsheel Park, New Delhi – 110 017, India
Penguin Group (NZ), 67 Apollo Drive, Rosedale, Auckland 0632, New Zealand
(a division of Pearson New Zealand Ltd)
Penguin Books (South Africa) (Pty) Ltd, 24 Sturdee Avenue, Rosebank, Johannesburg 2196, South Africa

Penguin Books Ltd, Registered Offices: 80 Strand, London WC2R 0RL, England

www.penguin.com

First published in Australia by the Text Publishing Company 2010
First published in Great Britain by Allen Lane 2011
Published in Penguin Books 2012
001

Copyright © Tim Flannery, 2010

Grateful acknowledgment is made for permission to reproduce the following illustrations:
Charles Darwin: University College London. The sand walk: Ted Grant.
Alfred Russel Wallace: The Wallace Fund, George Beccaloni. The island of Ternate: Tim Flannery.
James Lovelock: Bruno Comby. Tim Flannery: Nick Rowley. *Homo floresiensis* and *Elasmotherium sibiricum*:
Peter Schouten. Telefol elders and long-beaked echidna: Tim Flannery. Wisent: Romanowa.
Père David's deer: Lily M. Atomic bomb cloud: United States Department of Defense.
Greenland icecap: NASA. Attine ants: Arpingstone.

ISBN: 978-0-241-95073-9

www.greenpenguin.co.uk

MIX
Paper from
responsible sources
FSC™ C018179
www.fsc.org

Penguin Books is committed to a sustainable
future for our business, our readers and our planet.
This book is made from Forest Stewardship
Council™ certified paper.

ALWAYS LEARNING　　　　　　　**PEARSON**

To VJF, OMS and MH

Contents

A TURBULENT YOUTH

EVER SINCE AGRICULTURE

TOXIC CLIMAX?

or short and clean? The cost of unity and the gravest of crimes. The end of the global frontier. If we succeed much will be lost. Will it be Chinglish? Nature already ended? A domesticated Earth, or a re-wilded one? Paying the Medean debt. Fitting a brain to a body? The obligation of intelligence. The Faustus species. For the good of the Gaian whole. Earth as one entire, perfect living creature. Are we alone? A Universe to nurture the human spirit.

271

Foreword

This book is a twin biography of our species and our planet. At its heart lies an investigation of sustainability—not how we achieve it, but what it is. I have written it at a time when hope that humanity might act to save itself from a climatic catastrophe seems to be draining away. Yet I am not without hope, for I believe that as we come to know ourselves and our planet we will be moved to act. Indeed, provoking that action is the purpose of this book.

What is the nature of Earth? Is it akin to a cell, an organism or an ecosystem? How much energy does it require to operate? What is that energy used for, and how is it deployed? How flexible are Earth's systems? Can they withstand severe challenges, and can their resilience and productivity be enhanced?

And what of us? Are we constituted by natural selection to be so selfish and greedy that we're doomed to catastrophe? Or are there reasons to believe that we can overcome the problems confronting us, allowing our civilisation to continue? What of civilisation itself? What, precisely, is it?

These are some of the questions I attempt to answer in this book. Guiding me are the two great strands of evolutionary theory— reductionist science as epitomised by Charles Darwin and Richard Dawkins, and the great holistic analyses of the likes of Alfred Russel Wallace and James Lovelock. Each pursues a truth that at first seems to be in opposition to the other, but in the enormous complexity that is our living planet they operate as necessary and complementary opposites. When viewed together, these Darwinian and Wallacean world views, as I call them, provide a convincing explanation of life as a whole—and of what sustainability entails.

Fifty thousand years after our ancestors left Africa, our species is entering a new phase. We have formed a global civilisation of unprecedented might, a civilisation that is transforming our Earth. We have become masters of technology, spinning energy from matter at will and withal realising the dreams of the alchemists— transforming one element into another. We have trod the face of the Moon, touched the nethermost pit of the sea, and can link minds instantaneously across vast distances. But for all that, it's not so much our technology, but what we believe, that will determine our fate.

Today, many think that our civilisation is doomed to collapse. As I will show, such fatalism is misplaced. It derives in large part from a misreading of Darwin, and a misunderstanding of our evolved selves. Either such ideas will survive, or we will.

There are others who believe that endless growth is possible. In their imaginations only the fittest survive, and human intelligence will triumph over all. This optimism also derives from a misreading of Darwin, but it owes much as well to ignorance of the fundamentally important insights of Wallace and Lovelock. Despite their patently flawed nature, such foolishly optimistic ideas have now reigned largely unchallenged in western society for 150 years and have already led us far down the road to a dismal fate. Unless corrected, they may become a fatal flaw indeed.

Narrow horizons and short time frames are always misleading. That's why it's impossible to determine whether, even in the dramatic changes we see over a lifetime, we're witnesses to a descent into chaos, or a profound revolution that will lead to a better future. A wider view, one that encompasses humanity over the millennia and the world over the aeons, is required if we are to discern the true path of our evolutionary trajectory. In writing this book I've taken that long view, and, despite the challenges we now face, I feel optimistic—for ourselves, our children and our planet.

If we are to prosper, we must have hope, goodwill *and* understanding.

I

MOTHER NATURE
OR
MONSTER EARTH?

CHAPTER 1

Evolution's Motive Force

*There is nothing conscious about
life's lethal activities.*
PETER WARD 2009

Whatever each day held, Charles Darwin tried to set aside time for a stroll around a 'sand walk' near his home, Down House, in Kent. Tradition has it that the sand walk was his thinking space—the place where he sharpened his evolutionary theory, as well as the sentences that would so elegantly carry it into print. Consequently, the walk is regarded with reverence by many scientists, and when I made my first pilgrimage to Down House in October 2009 it was this place above all that I wished to see. After paying my respects to the great man's office and drawing room, I followed the signs to the walk. It's a little removed from the house and its enclosed gardens, and entering it one feels instantly transported from the ordered human world into the wider world of nature.

The walk consists of an oval-shaped path around a forest of hazel, privet and dogwood planted by Darwin himself. I was surprised to discover that despite its name there is no sand on it, nor has there ever been. Instead, it is surfaced with flints, which Darwin's son Francis remembered his father kicking from the path as a means of keeping count of the number of circuits he'd completed. The forest is now tall and venerable, and as I strolled I found myself pondering the thoughts that might possess a man as he walked repeatedly—almost compulsively—on a course as regular as a racetrack, through what must then have been saplings. While we can't know what occupied Darwin on the sand walk, there are hints in notes left by his children. As they grew up they took to playing in the walk, and often distracted and delighted their father with their games. To a man immersed in complex reasoning, such disturbances would surely be resented, so perhaps complex theories or elegant sentences weren't the things that occupied him after all.

It's my guess that during this repetitious physical activity Darwin was mentally fingering his worry beads—and looming large among his concerns were the implications of the theory he is now famous for. Known today as evolution by natural selection, it explains how species, including our own, are created. Natural selection, Darwin understood from his studies, is an unspeakably cruel and amoral process. He came to realise that he must eventually tell the world that we are spawned not from godly love, but evolutionary barbarity. What would the social implications be? As his discovery became widely understood, would faith, hope and charity perish? Would England's early industrial society, already barbaric enough, become a place where only the fittest survived, and where the survivors believed this was the natural order? Might his innocent-sounding theory turn people into cold-blooded survival machines?

Charles Robert Darwin was born in 1809 in Shropshire, the son of a wealthy society doctor. Baptised into the Anglican Church, he was expected to follow his father into medicine. But the cruelty of surgery in the pre-anaesthetic era horrified him, so he quit his studies in favour of training as an Anglican parson, and in 1828 he enrolled in a Bachelor of Arts degree at Cambridge. This was the (necessary) prerequisite for a specialised course in divinity, and in his finals he excelled in theology, while barely scraping through in mathematics, physics and the classics. Darwin's plans for a life of bucolic vicardom, however, were deferred when, in August 1831, he heard that a naturalist was needed for a two-year voyage to Tierra del Fuego and the East Indies aboard the survey ship *Beagle*.

Although his father initially opposed the venture, Charles won him over and was accepted as a self-funded gentleman naturalist on the voyage. His most important duty, from the navy's perspective, was to provide Captain Robert Fitzroy—a man of rather melancholy temperament—with companionship. The voyage would stretch to five years, taking Darwin round the globe and exposing him to the extraordinary biodiversity and geology of South America, Australia and many islands. It was in the Galápagos archipelago that Darwin collected what would become vital evidence for his theory—species of birds and reptiles that had evolved on, and were unique to, specific islands. For any young man such a voyage would be formative, but for Darwin it was world-changing. He later said that 'the voyage of the *Beagle* has been by far the most important event in my life and has determined my whole career'.

The experience led Darwin to reject religion. He later described how he had struggled to hold onto his faith, even as exposure to other cultures and the wider world made it less and less plausible:

> I was very unwilling to give up my belief; I feel sure of
> this, for I can well remember often and often inventing
> day-dreams of old letters between distinguished Romans,

and manuscripts being discovered at Pompeii or elsewhere, which confirmed in the most striking manner all that was written in the Gospels. But I found it more and more difficult, with free scope given to my imagination, to invent evidence which would suffice to convince me. Thus disbelief crept over me at very slow rate, but was at last complete.[1]

Upon returning to England in 1836, Darwin was accepted immediately into the bosom of the Victorian scientific establishment, and he commenced working up his *Beagle* discoveries. In 1842, aged thirty-two, he purchased Down House and there embarked upon a long career as an independent, and independently wealthy, scientist. The property provided for all Darwin's needs, serving as both a laboratory and a family home. Relatively modest in size, Down House must have been alive with the sounds of Charles and Emma Darwin's seven surviving children, and at times it must have seemed crowded. There is nonetheless an orderliness to the house and grounds that marks them as laboratories, in which Darwin pursued every conceivable ramification of the theory of evolution by natural selection, from the pollination of orchids to the origins of facial expressions.

Such a life is for the scientist a kind of Nirvana, but Darwin's lot was not entirely a happy one. Soon after returning from the *Beagle* voyage he fell ill, and for the rest of his life was plagued with symptoms, including heart palpitations, muscle spasms and nausea, that increased as he anticipated social occasions. Down House became his refuge, its solitude sustaining him through years of relentless work, illness and psychological stress until his death in 1882. I have little doubt that his illness was partly psychological, and exacerbated by what he believed to be the moral implications of his theory—a theory he largely kept to himself for twenty years.

Darwin had realised that new species arose by natural selection as early as 1838, but he didn't publish until 1858. 'It is like confessing a murder,' he confided to a fellow scientist when explaining his evolutionary ideas in a letter.

Down House is central to Darwin and the development of his theory, and to understand that extraordinary place one can do no better than to read Darwin's study of earthworms.[2] We might have earthworms in our gardens and compost bins, but few of us take the time to investigate them. For Darwin, however, they held a lifelong fascination. In many ways his worm monograph, which was his last book, is his most remarkable, documenting as it does experiments that ran continuously for almost three decades. Some of the worms lived in flowerpots, which were often kept inside Down House, and they seem to have become family pets. Certainly their individual personalities were appreciated, Darwin noting that some were timid and others brave, some neat and tidy while others were slovenly.

Eventually the entire Darwin family became involved in the worm experiments. I can picture Charles, surrounded by his children, playing the bassoon or piano to the worms in order to investigate their sense of hearing (they turned out to be entirely deaf), and testing their sense of smell (also alas rudimentary) by chewing tobacco and breathing on them, or introducing perfume into their pots. When Darwin realised that his worms disliked contact with cold, damp earth, he provided them with leaves with which to line their burrows, in the process discovering that they are expert practitioners of geometry (and indeed origami), for in order to drag and fold leaves efficiently, he noted, they must ascertain the shape of the leaf and grasp it appropriately. Darwin also provided his worms with glass beads, which they used to decorate their burrows in very pretty patterns. But, most importantly, he learned that worms profited from their experience, and that they were apt to be distracted from tasks by various stimuli he presented; and this, he believed, pointed to a surprising intelligence.

The sagacity and morality of worms were subjects Darwin never tired of. He concluded that wasps, and even fish such as pike, were far behind worms in their intelligence and ability to learn. Such conclusions, he said, 'will strike every one as very improbable', but:

> It may be well to remember how perfect the sense of touch becomes in a man when born blind and deaf, as are worms. If worms have the power of acquiring some notion, however rude, of the shape of an object and of their burrows, as seems to be the case, they deserve to be called intelligent, for they then act in nearly the same manner as would a man under similar circumstances.[3]

The worm monograph is also important in another way. In it Darwin came as close as he ever would to a sense of how Earth as a whole works. He had brushed against this subject in one of his early scientific papers that dealt with atmospheric dust he had collected while on the *Beagle*. Darwin thought that it was from the Sahara and was headed to South America, where the many spores and other living things included in it might perhaps find a new home. He never expanded his study into a theory of how dust might affect Earth overall, unlike more holistic thinkers we shall soon encounter who saw in dust important clues as to how life influences our atmosphere and climate. Darwin waited over half a lifetime before approaching what today is called Earth systems science—the holistic study of how our planet works—and, when he did so, it was through the lens provided by worms.

Darwin described how worms occur in great density over much

of England, and how they emerge in their countless thousands in the darkest hours, their tails firmly hooked in their burrow entrances, to feel about for leaves, dead animals and other detritus which they drag into their burrows. Through their digging and recycling they enrich pastures and fields, and so enhance food production, thereby laying the foundation for English society. And in the process they slowly bury and preserve relics of an England long past. Darwin examined entire Roman villas buried by worms, along with ancient abbeys, monuments and stones, all of which would have been destroyed had they remained at the surface; and he accurately estimated the rate at which this process occurs: about half a centimetre per year.

Darwin's monograph on worms reveals much of the man's temperament, and of his particular sense of humour. But it also highlights his strengths as a scientist—an ordered mind and immense patience. But patience can be a weakness too, and in the end it almost robbed Darwin of his future fame, for his dilatory approach to publishing the theory saw him nearly trumped by a man twenty years his junior, an unknown naturalist working in far-off Indonesia named Alfred Russel Wallace.

On 18 June 1858, Darwin received a letter from Wallace outlining a theory that described the way in which new species come into existence, and asking Darwin to transmit the manuscript to Charles Lyell, one of England's most eminent scientists, for publication. Darwin was devastated. 'I never saw a more striking co-incidence. If Wallace had my MS sketch written out in 1842 he could not have made a better short abstract,' wailed Darwin to his friend Lyell.[4] Only quick footwork by Lyell and another of Darwin's friends, the botanist Joseph Hooker, allowed Darwin's 'sketch' of 1842 and Wallace's paper to be published simultaneously by the Linnean Society of London on 1 July 1858.

As it was, neither Darwin's nor Wallace's paper attracted much immediate attention. In summarising the research published in the society's journal that year, President Thomas Bell was rather complimentary of the amount of botanical work completed, but lamented that the year 'had not, indeed, been marked by any of those striking discoveries which at once revolutionise, so to speak, the department of science on which they bear'.[5] To make an impression on the public, clearly something more was needed, and this Darwin produced the following year. On 24 November 1859 his book *On the Origin of Species by Means of Natural Selection, or the Preservation of Favoured Races in the Struggle for Life* was published. It was an instant success, forever securing Darwin's supremacy as the great evolutionist.

Despite being largely ignored, Darwin's first effort at introducing his idea got to the heart of the matter. In his 1858 paper he wrote:

> Can it be doubted, from the struggle each individual has to obtain subsistence, that any minute variation in structure, habits, or instincts, adapting that individual better to the new conditions, would tell upon its vigour and health? In the struggle it would have a better *chance* of surviving; and those of its offspring which inherited the variation, be it ever so slight, would also have a better *chance*. Yearly more are bred than can survive; the smallest grain in the balance, in the long run, must tell on which death shall fall, and which shall survive. Let this work of selection on the one hand, and death on the other, go on for a thousand generations, who will pretend to affirm that it would produce no effect?[6]

The essence of Darwin's insight is thus very simple. More are born than can survive, and those best fitted to the circumstances into which they are born are most likely to survive and breed. This selection of individuals, generation after generation, over the vastness of geological time, causes descendants to differ from their ancestors. There is no morality in this argument—no overall superiority of one individual, class or nation over another—for as the environment changes so do those selected as the 'fittest'. But it did reveal a terrible truth—the weak (poorly adapted) must die if evolution is to progress.

On that day in 1858 when his revolutionary idea was made known to the world, Darwin was unable to be with his assembled colleagues. He was instead mourning the death of his son, his namesake Charles. Always a frail child, Charles died of scarlet fever aged eighteen months. We can only imagine the mood in Down House that day. Infant death was far more common then, but not one whit less devastating. And the head of the family had just brilliantly elucidated the process that had rendered his child nothing but a cooling pile of flesh, food for worms. For Darwin, who believed that there was no hereafter and no God to comfort him in his grief, the blow must have been almost unbearable. And now he had to live with the thought that his theory might rob such comforts from the entire world.

It's hard to imagine, from today's perspective, the impact Darwin's book and theory had on society, but some sense of it can be gained from a debate held in Oxford's stately Zoology Museum in 1860. Arguing on Darwin's behalf was zoologist Thomas Huxley, later known as Darwin's bulldog, and opposing him was Samuel Wilberforce, the Bishop of Oxford, known as Soapy Sam on account of being one of the finest public speakers of his day. *On the Origin of Species* had been published just seven months earlier, splitting church and society. About a thousand people crowded between the skeletons, stuffed animals and mineral specimens to hear the bishop and the scientist slug it out. Hundreds more were turned away for lack of room, and Darwin, fast becoming a perpetual valetudinarian, was absent.

The critical moment came when Wilberforce took a cheap shot, asking whether Huxley was descended from an ape on his mother's or his father's side. This prompted an extraordinary response, which Alfred Newton, an eye-witness, described as follows:

> This gave Huxley the opportunity of saying that he would sooner claim kindred with an Ape than with a man like the Bishop who made so ill a use of his wonderful speaking powers to try and burke, by a display of authority, a free discussion on what was, or was not, a matter of truth, and reminded him that on questions of physical science 'authority' had always been bowled out by investigation, as witness astronomy and geology.
>
> He then got hold of the Bishop's assertions and showed how contrary they were to facts, and how he knew nothing about what he had been discoursing on.[7]

With the bishop embarrassed into silence, Admiral Robert Fitzroy, who had twenty-five years earlier been captain of the *Beagle* and Darwin's companion, rose to denounce Darwin's book and, 'lifting an immense Bible first with both hands and afterwards with one hand over his head, solemnly implored the audience to believe God rather than man'.[8] And there was the rub: Darwin, the erstwhile divinity student, was implying that ours is a Godless world, in which every kind of barbarity is condoned by nature.

Even today understanding of Darwin's theory remains mired in confusion and prejudice, and the mangled notions thus created have a malignant impact on society. Without doubt Darwin had settled upon an unfortunate subtitle for his work, for only upon reading the entire book would one discover that the 'favoured races' did not explicitly include the British ruling class. Almost immediately *On the Origin of Species* began to be used to justify the appalling social and economic inequalities of the Victorian era. The concept of the survival of the fittest was used to promote the notion that the misery of the poorest reflects the natural order. While Darwin must shoulder some of the blame for this, it's important to remember that it wasn't he who invented the term 'survival of the fittest', but the philosopher and libertarian Herbert Spencer, in 1864, who went on to apply Darwinian thought to his own social theories.[9] Darwin did however adopt the phrase in the fifth edition of *On the Origin of Species*, published in 1869.

There are other reasons for our partial failure to grasp Darwin's meaning, including religious and linguistic heritages. Nineteenth-century Christian dogma, with its insistence on literal creationism, survives into the twenty-first century, and although most mainstream religions have long accepted evolutionary theory (Darwin after all is buried in Westminster Abbey) opposition remains strong in some quarters. Just as importantly, the English language still lacks an easily understood term that elegantly conveys Darwin's insight. 'Evolution' hardly does the job. The word's Latin origins refer to the unrolling of a manuscript, and it's more of a magician's black box or cartoon caricature than an explanation to most people. Interestingly, Darwin himself hardly ever used the word, preferring 'descent with modification'.

Not all societies, however, are so handicapped. In 1898, the scholar Yan Fu translated Thomas Huxley's 1893 book *Evolution and Ethics* into Chinese. The Darwinian theories of human evolution expounded therein found ready acceptance in China, in part perhaps because they reflect some traditional Chinese folk beliefs about the stages of human development, which involve a progression from foraging, cave-dwelling ancestors to fire-using and house-building ones, and then to agricultural beings. In his translation, Yan Fu rendered the word 'evolution' as *tian yan*. Chinese characters can be read in several ways, and one way of reading these characters is as 'heavens' performance'—the heavens in this instance meaning all of creation.[10]

Yan Fu's phrase is now obscure and defunct, but heavens' performance strikes me as a beautiful and illuminating way of describing Darwin's discovery, for evolution is indeed a sort of performance, one whose theme is the electrochemical process we call life and whose stage is the entire Earth. Funded by the Sun, heavens' performance has been running for at least 3.5 billion years, and barring cosmic catastrophe will probably run for a billion more. It's an odd sort of performance, though, for there are no seats but on the stage itself, and the audience are also the players. Darwin's genius was to elucidate, with elegant simplicity, the rules by which the performance has unfolded.

One reason for the broad appeal of Darwin's ideas in the nineteenth and twentieth centuries is evident in the opening lines of his famous 1858 essay, with its reference to the Swiss botanist Augustin Pyrame de Candolle:

> De Candolle, in an eloquent passage, has declared that all nature is at war, one organism with another, or with external nature. Seeing the contented face of nature, this may at first well be doubted; but reflection will inevitably prove it to be true.[11]

War of course was one of the main occupations and passions of Victorian England, and the British excelled at it—the result being the greatest empire the world had ever seen. If nature favoured the militarily triumphant, then the Englishman must be a superior creature indeed. In an imperial age and aided by the works of

Spencer, Darwin's explanation of evolution would give rise to an extraordinary plethora of social phenomena, many of which strayed far from the original. Such beliefs are known as social Darwinism and, from colonial-era expressions like 'shouldering the white man's burden' and 'soothing the pillow of a dying race', and on to eugenics, they permeated the cultural and intellectual fabric of the era.

During the early part of the twentieth century the appeal of such thinking only strengthened. Indeed, by the 1930s and '40s social Darwinism was informing extermination and selective breeding programs in Nazi Germany, while in the US contributors to the journal *Eugenics* were arguing for the mass sterilisation of those they felt were inferior, as well as publishing ridiculous family pedigrees of the movement's leaders in an attempt to position them as the fathers of a future superior American race. Allied victory in World War II largely destroyed the credibility of these extremists and their programs, but some versions of social Darwinism continue to be influential. Notions about the 'survival of the fittest' are exemplified by Margaret Thatcher's comment in 1987 that 'there is no such thing as society' (by which she presumably meant that each should look after his own).[12] They are also evident in the field of neoclassical economics, with its belief that an unregulated market best serves humanity's interests.

Perhaps Charles Darwin, as he trod his sand walk, foresaw the possibility of all of this, or perhaps not. In any case, late in life he wrote, 'I feel no remorse for having committed any great sin but have often and often regretted that I have not done more direct good to my fellow-creatures.'[13]

CHAPTER 2

Of Genes, Mnemes and Destruction

*It would have been strange if philosophers
and naturalists had not been struck
by the similarity existing between the
reproduction in offspring...and that other
kind of reproduction we call memory.*
RICHARD SEMON 1921

Evolutionary theory has progressed enormously since Darwin's day, and without doubt the most important contribution has come from the discovery of the mechanism of inheritance—of genes, the structure of DNA, and genomes. The science that resulted from this fusion of Darwin's theory with genetics is called Neo-Darwinism, and its greatest exponent is Richard Dawkins. In *The Selfish Gene*, published in 1976, Dawkins outlines his thesis that the gene is the basic unit of natural selection. It has proved to be one of the most productive evolutionary insights ever, clarifying many aspects of Darwinian theory. In essence Dawkins argues that natural selection does not act primarily on us as whole organisms, but on each of the roughly twenty-three thousand genes that constitute the blueprint for our bodies. His work raises, in perhaps an even more acute manner than Darwin ever did, the moral dilemma that lies at the heart of Darwinism, for a central pillar of his reasoning is that we and other animals are mere 'survival machines' whose sole purpose is to ensure the perpetuation of the genes we carry.[14]

The defining quality of a successful gene, Dawkins believes, is 'ruthless selfishness'. In this he is a direct intellectual descendant of de Candolle, except that he believes the 'war' is being waged not only all around us, but inside our bodies as well. Indeed Dawkins' theory predicts that genes and the bodies they create are in competition. It explains, for example, why male spiders allow themselves to be eaten by females after mating (because it's good for the spider's genes), and why 'death genes' (that can kill individual organisms) exist in certain species. Musing on Tennyson's famous phrase, Dawkins says that 'I think "nature red in tooth and claw" sums up our modern understanding of natural selection admirably'.[15]

Dawkins has a genius for exposing the evolutionary mechanisms that lie hidden within us, and in doing so he highlights the limits of reductionist science in comprehending the complexity that is us. Consider his musings on maternal care:

> The sight of her child smiling, or the sound of her kitten purring, is rewarding to a mother, in the same sense as food in the stomach is rewarding to a rat in a maze. But once it becomes true that a sweet smile or a loud purr are rewarding, the child is in a position to use the smile or the purr in order to manipulate the parent, and gain more than its fair share of parental investment.[16]

It is not that this is wrong, only that Dawkins' mechanistic description of maternal love is inadequate to comprehend the profound relationship that exists between a mother and child. To thrive, a child must experience unconditional love, and a mother must feel that she is doing more than seeking rewards. No better illustration exists of why we cannot develop a satisfactory understanding of ourselves through selfish gene theory alone. We are far too complex to be comprehended through a reductive dissection of our parts.

We have a tendency to use ideas such as selfish gene theory to justify our own selfish and socially destructive practices. It's significant, I think, that Dawkins' book received wide acclaim on the eve of the 1980s—the era when greed was seen as good, and when the free market was worshipped. As our experience with social Darwinism illustrates, we need to be eternally on guard against the siren song of self-interest if we wish to live in a fair and equitable society.

Genes and ideas share at least one similarity: both reproduce, and the occasional error in reproduction provides variation. Thus, both are potentially subject to evolution by natural selection. Recognition that genes (or at least the physically inherited traits they give rise to) and ideas are similar is at least a century old. The German biologist Richard Semon wrote two books on the subject: *Die Mneme* (1904, published in English as *The Mneme* in 1921) and *Die Mnemischen Empfindungen* (1909, published in English as *Mnemic Psychology* in 1923).[17] He coined the word mneme (pronounced 'mnee-m', and which is derived from the Greek word for memory) to denote a grand unifying theory of reproduction— both physical and mental. He believed that memory had a physical reality, that it must leave an impression upon the brain. In describing his theory Semon wrote that:

> Instead of speaking of a factor of *memory*, a factor of *habit*, or a factor of *heredity*…I have preferred to consider these as manifestations of a common principle, which I shall call the *mnemic principle*.[18]

Semon's work catalogues a fascinating if all but forgotten episode in twentieth-century biology which sought to prove that experience could be inherited. He drew heavily on the work of Paul Kammerer, a brilliant young Viennese biologist whose experiments with what he called the fire-newt (*Salamandra maculosa*) were considered sensational at the time. Pregnant females were kept from water, thereby inducing them to give birth to fewer, more advanced young. This characteristic, it was claimed, was passed on to the next generation, despite their having free access to water. Other experiments, conducted by Marie von Chauvin on axolotls, resulted in the creatures developing lungs. Their offspring, she observed, frequently surfaced to gulp air, something normal axolotls will do 'only at an advanced age and in water deficient in air'.[19] But there was always the possibility that genetics, rather than Semon's 'mnemic principle', influenced the result.

Irrefutable proof, Semon felt, was at last obtained by the indefatigable Herr Kammerer. His triumph with the 'obstetric toad' (*Alytes obstetricans*) consisted of persuading the warty creatures to forgo having sex on land by keeping them 'in a room at high temperature…until they were induced…to cool themselves in the water-trough…Here the male and female found each other'. Forced to mate in water rather than on land, the toads coupled in a manner not usually favoured by the species.[20] This Semon interpreted as the creatures 'remembering' the ancestral method of copulation, which, it was claimed, persisted in subsequent generations.

Some of the experiments supposedly demonstrating the mnemic principle were truly bizarre. Dr Walter Finkler devoted himself to transplanting the heads of male insects onto females. The victims showed signs of life for several days but, perhaps unsurprisingly, exhibited disturbed sexual behaviour. Dr Hans Spemann made the 'Bombinator' frog grow eye lenses on the back of its head—a feat surpassed by Dr Gunnar Ekman, who induced green tree frogs (*Hyla arborea*) to grow eye lenses anywhere 'with the possible exception of the ear and nose primordia'.[21] This, Semon was convinced, demonstrated that frog skin 'remembered' how to grow eyes if appropriately stimulated.

By the 1920s the body of work Semon drew upon was under assault. The geneticists, championed by William Bateson (the originator of the term genetics itself), launched attacks that seem to have been vitriolic and obsessive. It has been suggested that Bateson had personal reasons for wishing to see Kammerer's work discredited, and when, in 1926, it was discovered that one of Kammerer's toads had been tampered with, this was held up as evidence that his entire body of work was suspect. With his reputation in tatters, Kammerer shot himself.[22]

Semon's all-encompassing theory did indeed have a fatal flaw: it necessitated a Lamarckian element in physical evolution. One of the iron-clad rules of physical evolution is that individuals cannot pass on to their offspring any favourable traits acquired during their lifetimes. Lamarck believed that giraffes could stretch their necks by continually reaching up for leaves, and that such stretched necks could be passed on to their offspring. Today we know that neck length among giraffes is coded in their genes, and that, with some rare exceptions (such as lengths of DNA inserted into genomes by viruses), physical traits acquired during an individual's lifetime cannot be passed on. Cultural evolution, in contrast, is purely Lamarckian. It is fuelled by the spread of ideas, and technologies that flow from such ideas, and those acquired by one generation are passed on to the next. Cultural evolution is far faster than physical evolution: it took the sabre-toothed cats millions of years to evolve their great stabbing canines, but it took humans only a few thousand years to develop metal daggers that are far more potent weapons.

For all its flaws, Semon's pioneering work held a seed of genius that is built upon in Richard Dawkins' book *The Selfish Gene*. Dawkins proposes the term 'meme' for transmitted ideas or beliefs. He says of them that, 'if memes in brains are analogous to genes they must be self-replicating brain structures, actual patterns of neuronal wiring-up that reconstitute themselves in one brain after another', adding that 'memes should be regarded as living structures, not just metaphorically but technically'.

In summary, Dawkins' memes are ideas that have a physical reality in our brains. They are transferrable just as genes are, and he suggests that they may be similarly selfish. Just how closely analogous mnemes (I prefer Semon's spelling) and genes are is an open question, but I do not believe that mnemes are necessarily selfish in the way that genes are. Some mnemes, for example, can see individuals act against their strict self-interest. Philanthropists often donate their wealth to causes that benefit humanity or the environment, and sometimes they do so anonymously, thereby ensuring that they accrue no social benefit. Perhaps they give to such causes simply because they believe it's the right thing to do. Whatever the case, such philanthropy is not in the interest of their selfish genes, which would benefit maximally if all was given to their children or near relatives.

Some mnemes, however, do prompt people to act selfishly, but such mnemes are decried in all societies. Indeed our strongest moral and religious precepts are aimed squarely at destroying them. As we've seen, such mnemes thrive at times, not least when given credibility by social Darwinism or Neo-Darwinian theory. Viewed in this light, the conflict between religion and evolutionary theory looks somewhat different. The challenge to religious belief that Darwinism presented in Victorian Britain acted as a kind of 'secret weapon' for the cause of selfish mnemes. By eroding religious authority it diminished, for some at least, a belief in the need for 'good works'. I find it interesting that our leading Neo-Darwinian, Richard Dawkins, is now engaged in a crusade against religion. Will this crusade leave in its wake a society in which it is more likely that ideas about selfish genes carry undue influence?

Selfish gene theory predicts that, in conflicts between genes and the bodies they create, genes will almost always prevail. But with the evolution of the mneme, all of that has changed. Humans have developed the idea (itself a mneme) of genetic engineering. The technology potentially allows us to snip genes we don't like out of our genomes. Clearly, in our modern age mnemes trump genes. Indeed mnemes are the most powerful things in the world. Around two hundred years ago a man called James Watt developed a mneme involving coal, steam and movement—and as a result the very composition of Earth's atmosphere today is changed.

It's often said that there are two fundamental sentiments that decide an election—hope for the future, and fear of it. If hope prevails, we're likely to elect more generous governments and reach out to the world, but if fear prevails, we elect inward-looking, nationalistic ones. Factors determining the successful spread of mnemes are clearly extremely complex, but at the broadest level it does seem that we, collectively and as individuals, gravitate towards one of these two tendencies. If we believe that we live in a dog-eat-dog world where only the fittest survive, we're likely to propagate very different mnemes from those that arise from an understanding of the fundamental interconnectedness of things. In large part, our future as a species will be determined by which of these mnemes prevails.

A reductionist view of evolution remains strong within the life sciences—and there's been a recent resurgence of interest in the power of Darwinian competition 'red in tooth and claw' to explain Earth history. The Medea hypothesis of palaeontologist

Peter Ward is named after the terrifying Medea of Greek mythology. Granddaughter of the Sun god Helios, she married Jason (he of the golden fleece), by whom she had two children. When Jason left her for Glauce, Medea extracted revenge by killing both of her children, after murdering Glauce and her father. Ward thinks that life is equally bloody and self-destructive, arguing that species will, if left unchecked, destroy themselves by exploiting their resources to the point of ecosystem collapse.[23] The Medea hypothesis in fact suggests that ruthless selfishness is inevitably a recipe for the elimination of a species. It argues that if we compete too successfully we will destroy ourselves.

Examples of Medean outcomes include the introduction of foxes into Australia in the nineteenth century, where they became so successful that they caused the extinction of twenty-odd native mammals that were their prey. If the settlers had not introduced rabbits, which the foxes also ate, the fox population would have suffered a catastrophic collapse. Easter Island offers another example. In this case humans destroyed the things their survival depended on, all the trees and birds, leading to population collapse and near extinction of people on the island.

Ward argues that the Medea hypothesis can explain the great extinction episodes of Earth's prehistory, and he sees the current destructive path of our human species as a continuation of that process. A key mechanism Ward identifies in causing these extinctions is the disruption of the carbon cycle by living things. One way this can occur is through what he calls 'greenhouse mass extinctions', which may be triggered if atmospheric carbon dioxide (CO_2) levels exceed a thousand parts per million. Essentially, Ward believes, the warming caused by CO_2 slows ocean circulation, depriving the ocean depths of oxygen. This allows sulphur bacteria (which don't need oxygen to live) to proliferate. Eventually oceanic oxygen levels drop so low that the sulphur bacteria reach the sunlit surface waters. There they release hydrogen sulphide (H_2S) into the atmosphere, destroying the ozone layer and poisoning life on land. With both sea and land devastated, up to 95 per cent of all species can become extinct, as happened 250 million years ago at the end of the Permian period.

But there are problems with this hypothesis, for it's not clear that the major extinctions which punctuate Earth's history were in fact caused by living things. Indeed some, such as the asteroid-induced extinction that carried off the dinosaurs, Ward acknowledges, were clearly not. Much more research on extinction events is required before Ward's hypothesis can be uncritically accepted. And of course it's important to understand that most species exist, most of the time, without destroying themselves or their ecosystems. But even if some extinctions—whether planetary or local—are caused by life itself, does that prove that we, Medea-like, are destined to destroy most other life, condemning our descendants to a new dark age, or outright extinction?

Perhaps the most important thing that the Medea hypothesis tells us is that Spencer's notion of the survival of the fittest should be turned on its head. If Ward is correct, then the fittest are merely engines of self-destruction, which through their success ultimately obliterate both themselves and most of the species they coexist with. Medea is also a deeply dismaying hypothesis, implying as it does that life has no choice: we must either thrive by destroying others or be destroyed ourselves in turn. In this manifestation, the Medea hypothesis represents a synthesis of Neo-Darwinism and an awareness of the limits and fragility of our environment.

Considering the deep contradictions between our popular ideas of survival of the fittest and Medean catastrophism, you could take the view that our belief systems are doomed to swing

incoherently between winner-takes-all theories of life and doomsday hypotheses. We will never understand our relationship to the
planet that is our home unless we sort through these contradictions.
But there has always been another approach, one that describes
the evolutionary process as a series of win-win outcomes that has
created a productive, stable and cooperative Earth—and its origins
can be found with the co-founder of evolutionary theory, Alfred
Russel Wallace.

CHAPTER 3

Evolution's Legacy

He who opens his eyes to the possibilities of
evolution in their endless variety will abhor
fraud and violence and disdain prosperity
at the expense of his fellow creatures.
SVANTE ARRHENIUS 1909

Despite the fact that Charles Darwin and Alfred Russel Wallace independently hit upon the theory of evolution by natural selection, two more different men there never were. Darwin was a patient, methodical toiler, a scientist in the finest reductionist tradition. Wallace, in contrast, was a great synthesiser of all he saw and sensed, whose ideas came as flashes of genius. His description of the evolutionary process was dashed off in a few hours while he was in the grip of a malarial fever on the island of Ternate, in what is now Indonesia, yet it is the intellectual equal to Darwin's painstaking effort. In summarising his theory Wallace said:

> There is a tendency in nature to the continued progression
> of certain classes of varieties further and further from the
> original type...This progression, by minute steps, in various
> directions, but always checked and balanced by the necessary

conditions, subject to which alone existence can be preserved, may, it is believed, be followed out so as to agree with all the phenomena presented by organised beings, their extinction and succession in past ages, and all the extraordinary modifications of form, instinct, and habits which they exhibit.[24]

Darwin could hardly have put it better, but it's what Wallace did with his life after 1858 that sets him apart—for while Darwin sought enlightenment by studying smaller and smaller pieces of life's puzzle, Wallace took on the whole, trying to make sense of life at a planetary and universal scale. As he aged Wallace would, I think, have increasingly appreciated Yan Fu's translation of evolution as heavens' performance.

Born in Wales in 1823, Wallace was a product of the working class, and he embodies as much as anyone the aspiration for self-improvement characteristic of the age. Withdrawn from school because his family was unable to pay the fees, he joined his elder brother as an apprentice builder before an economic downturn left him briefly unemployed. Then, in 1848, he set off for Brazil to work as a collector of natural history specimens. In this he was fantastically successful, but as he was returning home laden with butterflies, birds and beetles sufficient to set him up for life, disaster struck. It began with the captain of the vessel that was carrying him and his collection emerging on deck. 'I'm afraid the ship's on fire,' he said. 'Come and see what you think of it.'[25] There was little time to think, however. The hold was filled with highly flammable palm oil, and Wallace could only grab a box containing a few drawings, clothes and one diary, before leaping into a lifeboat. Everything else, including his extensive scientific notes and magnificent collections, was lost. After ten days adrift in the mid-Atlantic the survivors were picked up by another vessel, but its provisions were exhausted, so the men were reduced to catching the ship's rats for food, and even consumed the contents of the grease pot. Just when it seemed that things could get no worse, another disaster struck the emaciated, rag-clad survivors. When almost in the English Channel, the ship was caught in a tremendous gale and, by the time it limped into London, sea water more than a metre deep had collected in the hold. Escaping a second shipwreck by a hair's breadth, Wallace arrived home penniless and looking like a castaway. He was soon forced by economic necessity to return to the tropics. This time he went to the East Indies, where he would remain until 1862, amassing collections and discoveries sufficient for enduring fame.

If Darwin was at the centre of the scientific establishment, Wallace was perpetually on its margin. Self-educated and perhaps not sceptical enough in some matters, he infamously became the dupe of spiritualists, his patronage giving credibility to their sleight-of-hand tricks that promised to put people in touch with the dead. He vehemently opposed vaccination on the eminently sensible grounds that there were risks involved in transmitting bodily matter between species and individuals. But he failed to see that even in that age of rudimentary hygiene the benefits of inoculation far outweighed the risks, and the medical establishment excoriated him for slowing vaccination's public acceptance. Later in life Wallace came to believe that the raison d'etre of the Universe was the development of the human spirit, a view widely mocked as naïve and anthropocentric.

All of this was sufficient reason for the Victorian elite to exclude this self-made scientist, but Wallace was objectionable for another reason, one which struck at the very source of their wealth. One of his chief preoccupations was the air pollution then choking Britain's cities. He believed that 'the vast manufacturing towns belching forth smoke and poisonous gases' were stunting the bodies of working-class children, and indeed they were, carrying countless thousands prematurely to their graves.[26] An activist for social justice all his days, Wallace argued that this pollution persisted because of 'criminal apathy'.

Wallace lived to be ninety and, as he aged, his mind turned increasingly to the question of how the Earth as a whole works. *Man's Place in the Universe*—one of his last books, published in 1904, when he was eighty—has as its principal objective to demonstrate that life is unique to Earth, other planets such as Venus and Mars being dead.[27] It is perhaps the foundation text of astrobiology. Wallace elucidates the importance of the atmosphere to life in chapters such as 'Clouds, Their Importance and Their Causes' and 'Clouds and Rain Depend upon Atmospheric Dust'. And it is in this seemingly trivial matter of atmospheric dust that we see a difference between Darwin and Wallace: for Darwin it remained merely a phenomenon of zoogeographic interest, helping to explain the distribution of microorganisms, while Wallace saw it as an absolutely essential element in the Earth system, responsible for the blessings of rain and clouds, and, as such, deeply influential on the entire planet's climate. In describing the atmosphere as a whole, Wallace said:

> It is really a most complex structure, a wonderful piece of machinery, as it were, which in its various component gases, its actions and reactions upon the water and the land, its production of electrical discharges, and its furnishing the elements from which the whole fabric of life is composed and perpetually renewed, may be truly considered to be the very source and foundation of life itself.[28]

Unlike Darwin, Wallace seems to have had no fear that an understanding of evolution would corrupt public morality—indeed he saw the evolutionary process, and our understanding of it, as potentially ushering in a wonderful future. I think that's because Wallace realised that while evolution by natural selection is a fearsome mechanism, it has nevertheless created a living, working planet, which includes us, with our love for each other, and our society. When I look out of the windows of my house near Sydney I can see the world of Wallace's vision. It is manifest in a graceful, pink-barked angophora tree that spreads a bounteous shade—a tree composed of billions of individual cells. Once, the ancestors of the chloroplasts that give its leaves their green colour were free-living bacteria. Then, aeons ago, they came to live within a single-celled, primitive plant. Today, so complete is the union of these once free-living and only remotely related organisms that most of us think of them as one, in this case a tree.

There is a more modest tree nearby called the scribbly gum, a whitish, twisted thing that bears an indecipherable script written by a beetle on its bark. The beetle cannot live without the tree, and the tree cannot live without an invisible partner, a fungus so humble that it cannot be seen, which sheaths intimately the scribbly gum's finest rootlets and improves the tree's access to nutrients. Fungus, beetle, bird, tree, and the human sitting in its shade, joyed by the song of the bird and the thought that a beetle has learned to write on bark. We are part of an interdependent community.

And then there is me. Billions of cells cooperating seamlessly at every moment and a brain made up of a reptilian stem, a middle mammalian portion, and two highly evolved yet relatively poorly connected hemispheres somehow add up to that thing I call me. And beyond that miracle of cooperation is my wider world, made up of a web of loves that I could not live without: spouse, children, parents, friends. Who is to say that a marriage cannot be any less complete a union than that between a chloroplast and the cell that hosts it? Beyond my family circle there is my city with its millions of residents, my country, which coordinates actions through a ballot box, and beyond that my planet with its countless dependent parts. Our world is a web of interdependencies woven so tightly it sometimes becomes love.

There are no doubt people who believe this cannot be, who argue that they inhabit a world ruled by intense competition in every domain of life, and that any semblance of love for our fellow man results only from God's good grace. True enough, competition exists, but it is 'the contented face of nature' that Darwin wrote of so sceptically that reigns most of the time. And from the love that sustains my family to the beetle that writes on the tree, every bit stems from evolution by natural selection.

If competition is evolution's motive force, then the cooperative world is its legacy. And legacies are important, for they can endure long after the force that created them ceases to be.

A clear illustration of the process that has created life as we know it can be found in a game of football. Anyone reading the sports pages might think that football is all about competition, but you've only got to see a match to learn how wrong that is. Football is a miracle of cooperation, and it's not just the teams that exist in extraordinary wholeness for that brief span between kickoff and final whistle. The eruption of emotion at a goal and the hushed silence at a last-minute free kick reveal a union of feeling in the spectators that lies at the very heart of the sport. After all, it's the sense of being part of this greater whole that gets the fans to the game each week, and without them the game would not exist. In sport, winners can survive only if losers do too; otherwise, there'd be no game. Our planet is rather similar. If a sufficiently superior and arrogant species arose and pursued a winner-take-all philosophy, it would be game over for us all. Alfred Russel Wallace, I believe, was the first modern scientist to comprehend how essential cooperation is to our survival.

I sometimes ponder what our world would be like if Wallace, rather than Darwin, had become the great scientific hero of the age. Would evolutionary theory have become the justification for an unjust society? Would evolution instead have been harnessed to an agenda of social reform? Would the sciences of ecology and astrobiology have emerged a century before they in fact did? Would air pollution and climate change have been defeated in the nineteenth century? We shall never know the answers to these questions. With the exception of a few, such as the Nobel Prize-winning chemist Svante Arrhenius, the scientific mainstream resisted most of Wallace's ideas. Ironically, he's best remembered for his zoogeography—the Wallace Line, a boundary that separates animal species in Australia and New Guinea from those in Asia.

Wallace was a profound thinker, yet his deepest ideas could not prosper in the brutal, imperial age in which he lived. But times change, and when in the 1970s a more powerfully explicatory theory of the type Wallace was groping at emerged, the world was at last prepared to listen.

The person who developed that theory was James Lovelock, and he did so, as far as I can determine, without knowledge of Wallace's work. Indeed, it's a remarkable fact that most of the researchers working in what we might call the Wallacean tradition of holistic, planetary-scale science seem to have arrived in the field more or less independently, unaware of the writings of their predecessors. Perhaps this is because the Wallaceans have rarely been part of the academic mainstream. Whatever the case, Wallace and Lovelock were both working-class outsiders of exceptional ability, and both saw that the atmosphere was the key to understanding life as a whole.

James Lovelock was born in Letchworth, outside London, in July 1919—a result, he believes, of the Armistice celebrations in November 1918. Although an indifferent student, by the age of twelve he had determined to become a scientist and began frequenting public libraries. James Jeans' *Astronomy and Cosmogony*, Frederick Soddy's *The Interpretation of Radium* and L. G. Wade's *Organic Chemistry* became his most cherished reading. It was at this time that he drifted from his agnostic upbringing and for a while became a Quaker with strong pacifist views. He studied chemistry at Manchester University, and, in 1941, got a job at the National Institute for Medical Research, where one of his principal responsibilities was investigating air hygiene in bomb shelters. An inventor of scientific instruments, he produced several devices to measure atmospheric composition, and so began a romance with the atmosphere that would last a lifetime.[29]

Lovelock tells us that the concept of Gaia came to him suddenly one afternoon in September 1965, when he was visiting the Jet Propulsion Laboratory in California. An astronomer had brought him data gathered by infrared-detecting instruments from the atmospheres of Mars and Venus, which revealed for the first time that they were composed principally of CO_2. Lovelock realised immediately that this was evidence that both Venus and Mars were dead planets, and that Earth was different because living things had reduced its atmospheric CO_2 and replaced it with oxygen. When he mentioned this to the American astrophysicist Carl Sagan, Sagan told him of the 'faint young Sun paradox', which states that, while the Sun was 25 per cent cooler three billion years ago than it is today, our planet never froze right over as it would seem it should have. Then, Lovelock says, 'the image of the Earth as a living organism able to regulate its temperature and chemistry at a comfortable steady state emerged in my mind'.[30]

The Gaia hypothesis has gained a reputation as being somewhat 'new age', as superficial popular science. However, it is anything but, being soundly based and profoundly important to our understanding of the evolution of life on Earth. In universities it is often studied as 'Earth systems science', perhaps because that sounds more respectable. Today Lovelock describes Gaia as:

> A view of the Earth...as a self-regulating system made up from the totality of organisms, the surface rocks, the ocean and the atmosphere tightly coupled as an evolving system...this system [has] a goal—the regulation of surface conditions so as always to be as favourable as possible for contemporary life.[31]

When an account of the Gaia hypothesis was first published in 1972 in the journal *Atmospheric Environment* it gained little credibility among scientists.[32] The situation did not improve after Lovelock published his book *Gaia* in 1979. 'The biologists were the worst,' he recalls. 'They spoke against Gaia with the kind of dogmatic certainty I hadn't heard since Sunday School. At least the geologists offered criticisms based on their interpretation of the facts.' Some of the most important criticisms came from Richard Dawkins, who described Lovelock's book as part of 'the pop-ecology literature'.[33] The hypothesis, Dawkins believed, did not take proper account of evolution by natural selection, with its requirement for competition between organisms, writing that:

> There would have to have been a set of rival Gaias, presumably on different planets. Biospheres which did not develop efficient homeostatic regulation of their planetary atmospheres tended to go extinct. The Universe would have to be full of dead planets whose homeostatic regulation systems had failed, with, dotted around, a handful of successful,

well-regulated planets of which Earth is one...In addition
we would have to postulate some kind of reproduction,
whereby successful planets spawned copies of their life forms
on new planets.[34]

These criticisms prompted Lovelock to investigate how a
process based on competition could create 'the contented face of
nature' and to do so he developed a computer model, in 1982,
known as Daisyworld.

Daisyworld is an attempt to see what would happen on an
imaginary planet with a very simple ecology that followed the
same orbit around the Sun as Earth. Only daisies grow there,
and they vary from dark- to light-coloured. They can grow only
in a temperature range of -5° to 40° Celsius, with an optimum
temperature of 20° Celsius. The only thing that affects the temperature of this model world is how reflective its surface is: if it's bright
then more sunlight is reflected into space before it turns into
heat energy; if dark, then lots of sunlight is turned into heat energy
and so Daisyworld heats up. A bright daisy will thus cool its
surroundings, while a dark one will warm them.

In order to investigate the 'faint young Sun paradox', Lovelock
ran programs to simulate conditions as they have been throughout
Earth's history. As the programs ran, large clusters of light-coloured
daisies died off because their surroundings became too cool, while
similar clumps of dark ones died because they became too warm.
Over numerous computer generations of daisies, the proportion of
light and dark types became balanced so as to keep conditions at
the surface relatively constant and within the optimum temperature
range for daisy growth. Over the years, more complex Daisyworld
models have been developed that better mimic the natural world.
But always the results are the same: life as a whole (albeit virtual

life) regulates conditions to suit itself. That is, until it meets a force so great—such as an asteroid or emission of greenhouse gas—as to overwhelm its control mechanisms.

Calling Daisyworld his 'proudest scientific achievement', Lovelock argues that it completely answers the criticism that Gaia could not evolve by the process of Darwinian natural selection. It's a view championed by Mark Staley, one of the foremost proponents of Daisyworld-type modelling, who says of the models that 'the end result may appear to be the product of a cooperative venture, but it is in fact the outcome of Darwinian selection acting upon "selfish" organisms'.[35] Several real-world examples of Daisyworld-like regulation have now been discovered. Among the most intriguing is the way coral reefs increase cloudiness in the air above them through the production of cloud-seeding chemicals, thus shading themselves from dangerous ultraviolet radiation. Another example concerns rainforests like the Amazon, which transpire water vapour, generating their own rainfall.

In summary, Lovelock's Gaia hypothesis describes cooperation at the highest level—the sum of unconscious cooperation of all life that has given form to our living Earth. It's not that living things choose to cooperate, but that evolution has shaped them to do so. It also shows that the living and non-living parts of Earth are inextricably interwoven. Lovelock argues, for example, that 99 per cent of Earth's atmosphere is a creation of life (the exceptional 1 per cent being the noble gases such as Argon) and that Earth's oceans are maintained in their current state by life itself. But most importantly, the Gaia hypothesis posits that Earth, taken as a whole, possesses many of the qualities of a living thing.

It was the novelist William Golding who suggested the name Gaia for Lovelock's hypothesis. Golding then lived in the same village as Lovelock, and doubtless knew of Gaia, the Greek goddess of Earth, from his readings of the classics. Perhaps it required the author who in 1954 had written *Lord of the Flies*, arguably the most terrifying 'survival of the fittest' novel ever published, to provide the modern world with a name for a unified theory of life. Two decades later Golding returned to the contemplation of Gaia. In a 1976 review in the *Guardian* of a book of aerial photography, he opined:

> Our growing knowledge both of the microscopic and the *macroscopic* nature of Earth is not just a satisfaction to a handful of scientists. In both directions, it is bringing about a change in sensibility...Those who think of the world as a lifeless lump would do well to watch out.[36]

The idea of Earth as a living entity is not new. In explaining the ancient Greeks' thoughts about the Earth, Sir Francis Bacon wrote in 1639 that:

> The philosophie of Pythagoras...did first plant a monstrous imagination, which afterwards was, by the school of Plato and others, watered and nourished. It was, that the world was one entire perfect, living creature...This foundation being laid, they might build upon it what they would; for in a living creature, though never so great, as for example a great whale, the sense and the effects of any one part of the body instantly make a transcursion throughout the whole.[37]

Hailed as one of the founding fathers of modern science, Bacon was also a deeply religious man, and his revulsion at the Greek concept of Earth as a living thing stems in part from the church's battle with witchcraft, which was waxing hot in seventeenth-century England. If the Earth was 'one perfect, living creature', then Bacon felt that witches and sorcerers would be able to influence any part of it from a distance, just as a tweak of the toe can make an entire body jump. Their satanic works were, he feared, the deft touch to the Earth-body that might conjure storms to destroy ships at sea, or incite earthquakes to crush the cities of the righteous.

But Christian antagonism to the idea of a living Earth goes far deeper than that. After all, Gaia is a pagan god, and the early church waged a fierce battle with such competitors. It largely succeeded in imposing monotheism in western Europe, and by the eighteenth century belief in the Earth as anything like a living thing survived only in the minds of the simplest and most unschooled of peasants. In the churches and universities, in contrast, Earth was seen as a stage upon which the great moral drama of good and evil was being played out, at the end of which we would be consigned to either heaven or hell. And it was a stage over which we had been granted dominion, to treat as we liked—a view that the magnates of the industrial revolution would exploit for their own ends.

Lovelock's hypothesis is at least as controversial today as Darwin's theory of evolution was 150 years ago. Part of the reason can be found in its history of conflict with Christianity. There are still church leaders who denounce environmentalism as if it somehow competes with their version of religious dogma. Australia's leading Catholic, Cardinal George Pell, believes that environmentalists suffer from a new 'pagan emptiness'. Even worse, from Pell's perspective, they compete with religion. In January 2008 he said of climate science:

> The public generally seem to have embraced even the wilder
> claims about man-made climate change as if they constituted
> a new religion. These days, for any public figure to question
> the basis of what amounts to a green fundamentalist faith is
> tantamount to heresy.[38]

And of course the deep interconnectedness central to the Gaia hypothesis presents a profound challenge to our current economic model, for it explains that there are both limits to growth, and no 'away' to throw anything to.

Within mainstream science the Gaia hypothesis long remained marginal: neither Wallace nor Lovelock ever held a university position, and there has not, until recently, been a Wallacean academic tradition. But slowly that is changing. Within geology, Earth systems science is finding some respectability, and even in the biological sciences academics are turning their attention to Gaian questions, including how the evolutionary process might have fostered cooperation. Those interested in such questions are known as sociobiologists. Oxford University's Bill Hamilton is widely seen as the discipline's founder, and Harvard's E. O. Wilson its greatest living exponent. Sociobiology is a synthetic science which seeks to explain the social behaviour of animals through evolutionary theory. Some scientists (among them Stephen Jay Gould) have aligned sociobiology with social Darwinism, but in reality it deserves to be classified with the other synthetic sciences, such as astrobiology and Earth systems science. As we explore the themes of this book we'll hear more of it, and particularly of its founder, the great Bill Hamilton, for he came as close as anyone ever has to bridging the gulf between Neo-Darwinism and the Gaia hypothesis.

CHAPTER 4

A Fresh Look at Earth

*Big ball of iron with some rock on the outside
and a very very thin coating of moisture
and oxygen and dangerous creatures.*
A DESCRIPTION OF EARTH, WIKIPEDIA

What is life? Is it separable from Earth? At the most elemental level, we living beings are not even properly things, but rather processes. A dead creature is in every respect identical to a live one, except that the electrochemical processes that motivate it have ceased. Life is a performance—heavens' performance—which is fed and held in place, and eventually extinguished, by fundamental laws of chemistry and physics. Another way of thinking about life is that we are all self-choreographed extravaganzas of electrochemical reaction, and it is in the combined impacts of those reactions, across all of life, that Gaia itself is forged.

Thinking of life as something separate from Earth is wrong. A striking instance concerns the origins of diamonds. Analysis shows that many diamonds are made from living things. Tiny organisms adrift on an ancient sea took in carbon from the atmosphere, then died and sank into the abyss. From there geological processes carried the carbon into the Earth's very mantle, subjecting it to unimaginable heat and pressure, thereby transforming it into diamonds. Eventually these were shot back to the surface in great pipes of molten rock, and today some grace our fingers.[39]

Our planet formed some 4.5 billion years ago as a result of a 'gravitational instability in a condensed galactic cloud of dust and gas'.[40] It formed in an astonishingly short time, perhaps as little as ten million years, and critically important qualities were added when a heavenly body the size of Mars struck the proto-Earth, liquefying it and ejecting from it a mass destined to become the Moon. The liquefied remainder then began to differentiate into a metallic core, making up almost 30 per cent, a silicate mantle making up almost 70 per cent, and a thin crust making up just 0.5 per cent. Within a billion years, or perhaps just a few hundred million years, parts of that crust had begun to organise into life.

That was so long ago that the Moon was far closer than it is today, and was replete with active volcanoes. It loomed large in the sky, and exerted such gravitational pull that Earth's crust buckled many metres with each tidal swing. It challenges our imagination to think of microscopic portions in that ancient crust slowly becoming living things, and indeed how the spark of life was first kindled remains one of science's great mysteries. But there is no doubt that the electrochemical processes that are life are entirely consistent with an origin in Earth's crust—our very chemistry tells us that we are, in all probability, of it. This concept of life as living Earthly crust challenges the dignity of some. It should not. We have long understood, from biblical teaching and practical experience, that we are naught but earth: ashes to ashes, dust to dust, as the English burial service puts it. Indeed, 'dust thou art, and unto dust shalt thou return' are among the oldest written words we have.[41]

The building blocks of life, however, go back even further

than the formation of our planet. The elements that form us, the carbon, phosphorus, calcium and iron to name but a few, were created in the hearts of stars. And not just in one generation of stars, for it takes the energy of three stellar generations combined to form some of the heavier elements, such as carbon, that life finds indispensible. Stars age very slowly, and to complete three generations takes almost all of time—from the Big Bang to the formation of Earth. We are, as the astrophysicist Carl Sagan said, mere stardust, but what a wondrous thing that is.

Earth's crust may seem like a passive organ, a mere substrate, but it has been profoundly influenced by life, and it is the sheer size of life's energy budget (the total amount of energy living things capture from the Sun) that makes this possible. Plants capture the Sun's energy using photosynthesis. Inside green leaves lie tiny structures, called chloroplasts, which use the energy of sunlight to break apart molecules of CO_2 which, if they were not so dealt with, would eventually make up most of Earth's atmosphere. Plants use the CO_2 to form organic compounds, which in turn are used to build bark, wood and leaves—indeed all the tissues of the plants around us. Look at a tree and what you see is mostly congealed carbon, a tonne of dry wood being the result of the destruction, by photosynthesis, of around two tonnes of atmospheric CO_2.

Green plants are far more efficient in their energy use than we humans with our fossil-fuelled power stations. Each year green plants manage to convert around one hundred billion tonnes of atmospheric carbon into living plant tissue, and in so doing they remove 8 per cent of all atmospheric CO_2. This is a truly extraordinary figure. Just imagine if no CO_2 found its way into the atmosphere. In just twelve years plants would then absorb and use almost *all* of the atmospheric CO_2.

Plants capture about 4 per cent of the sunlight that falls on Earth's surface, which gives life a primary energy budget (excluding sulphur bacteria and other non-photosynthetic pathways) of approximately one hundred terawatts (one hundred trillion watts) annually. It's the size of life's primary energy budget and the resilience of its ecosystems (which is determined in part by biodiversity) that define a healthy planet. Scientists have only begun to think about Earth in these terms, so measurements of productivity and diversity remain approximate. Yet it's clear from major extinction events in the fossil record that if Earth's energy budget and ecosystem resilience fall below certain thresholds, a fully functioning Earth system cannot be maintained.

Useful parallels can be drawn between the way energy flows in economies and Earth's ecosystems. The size of economies is measured in dollars, while Earth's energy budget is measured in terawatts. Dollars and terawatts clearly differ, but both represent potential resources that can be put to productive ends. Although an area of active study and dispute, it seems that the stability of both economies and ecosystems is related to their diversity, which itself is partly a function of size: the larger an economy or an ecosystem, the more diverse it can be. The presence of certain elements in economies and ecosystems can also help foster productivity. Banking is a good example. In economies well-run and well-regulated banks aid the flow of capital, thus stimulating productivity. In ecosystems certain species act rather like bankers by facilitating energy and nutrient flows. Earth's ecological bankers include the big herbivores, those weighing a tonne or more. As we'll soon see, in marginal ecosystems such as deserts or tundra, these ecological bankers speed the flow of resources through the ecosystem, allowing a substantial 'biological economy' to be built on a slender resource base. If humans destroy megafauna, they can induce the equivalent of a never-ending recession on such ecosystems, limiting their productivity and stability. And that impacts Earth function as a whole, just as a recession in the US can affect the global economy.

So how does life spend its capacious energy budget? Basically, it is deployed to modify our planet so as to make it more habitable, and just how that is done is best understood by comparing Earth with the dead planets, such as Venus and Mars. Planets can have up to three principal 'organs', which correspond to the three phases of matter: a solid crust, a liquid (or frozen) ocean and a gaseous atmosphere. A living planet uses its energy budget to kick the chemistry of its organs out of balance with each other. No greater example of this exists than oxygen. Earth's atmosphere is full of this highly reactive element, but if life was ever extinguished oxygen would quickly vanish by combining with elements in the rocks and oceans, forming molecules such as CO_2. The chemical composition of the organs of dead planets, in contrast, exists in a state of equilibrium. As Lovelock realised in the 1970s, a planet whose atmosphere consists almost entirely of CO_2 is a planet whose life force, if there ever was one, is long exhausted—a planet at eternal rest.

Carbon is the indispensible building block of life. You and I are made up of 18 per cent carbon by dry weight, and plants have a much higher percentage. Almost all of that carbon was once floating in the atmosphere, joined in a *ménage à trois* with oxygen to form CO_2. Billions of years ago, when life was a weak infant struggling to survive, there was more CO_2 in the atmosphere than there is today, for living things had not yet discovered a means to use it. Back then, perhaps, life nestled as microscopic bacteria in the bosom of the deep sea, or hid in sediments around hot springs. Wherever it found a refuge, its energy budget must have been small, as most of Earth was still untouched by its power. Today, however, CO_2 forms just four parts per ten thousand of the gaseous composition of Earth's atmosphere, while a by-product of photosynthesis, oxygen, forms 21 per cent. This is the ultimate measure of life's triumph.

Earth's continental crust is far thicker than its oceanic crust and it's made of lighter, silica-rich rock. The continents originated from erosion of the oceanic crust (which is made of basalt) and, remarkably, they may be a product of life. This might seem to be a large claim, but it's worth keeping in mind that living things provide 75 per cent of the energy used to transform Earth's rocks, while heat from within the Earth provides a mere 25 per cent.[42] We tend to think about the transformation of rocks in Earth's crust as the result of volcanoes, earthquakes and such like. It's easy to overlook the silent work of lichens, bacteria and plants, which create grains of soil from intransigent basalt and other rocks by reaching deep into the strata, leaching and breaking down the rock with the acids they exude. Their work, while microscopic in scale, is ceaseless, and thrice greater in effect than that of all the world's volcanoes combined.

We have no evidence of life for the first half-billion years or so of Earth's existence. Back then our planet was a water-covered sphere with little or no dry land. When life originated, those ancient living things, it has been suggested, produced acids that sped up the weathering process of the basaltic crust, separating the lighter elements in the basalt from the heavier ones. When these lighter elements are compressed and heated by movements in Earth's crust they become granite, the foundation-stone of the continents and the essence of the earth beneath our feet. Perhaps, given enough time, energy from within the Earth could have affected the same transformation, but so vast was the amount of basalt weathered to create the first continents that recent research indicates it could have occurred only if life was capturing energy and using it to produce compounds that help break down rocks.[43]

We can think of Earth's rocky crust as a huge holdfast, like the lower shell of an oyster, which life has formed to anchor itself. And if we imagine the rocks as life's holdfast, then we can think of the atmosphere as a silken cocoon, woven by life for its own protection and nourishment. Just consider what the atmosphere does for us. Its greenhouse gases keep the surface of the planet at an average of around 15° Celsius, rather than -18° Celsius. All of the principal greenhouse gases are produced by life (though some, such as CO_2, can be produced in other ways as well), and without them Earth would be a frozen ball. Ozone, a form of oxygen composed of three atoms bonded together, is a product of life, for all free oxygen is derived from plants. While it makes up just ten parts per million of our atmosphere, it captures 97 to 99 per cent of all ultraviolet radiation heading our way. Without this protection, our DNA and other cellular structures would soon be torn apart and life at Earth's surface would cease to exist. Then there is the more common form of oxygen (two atoms bonded together), which fuels our inner metabolic fires, providing the breath of life itself.

As Wallace knew, our atmosphere is truly wondrous. We may think of it as big, but it is by far Earth's smallest organ. To compare it with the oceans, we need to imagine compressing its gases around eight hundred times, until it becomes liquid. If we could do that, we'd see that the atmosphere is just one-five hundredth the size of the oceans. It's a delicate, dynamic and indispensible wrapping to the planet, a cocoon that is constantly being repaired and made whole by life itself, a cocoon that intimately wraps around every living thing and connects chemically with a great rocky shell that life has forged as its support. And sandwiched between holdfast and cocoon is the liquid circulatory system of the beast: Earth's oceans and other waters.

Earth is truly the water planet, for water in its three states—

vapour, liquid and solid—defines and sustains it. Liquid water covers 71 per cent of Earth's surface while solid water, mostly in the form of glacial ice, covers a further 10.4 per cent. Water is essential to life because the electrochemical processes that are life can occur only within it; fluids as salty as the ancient oceans flow through our veins. The ocean was almost certainly the cradle of life, and it remains life's most expansive habitat. With a volume of 1.37 billion cubic kilometres it is eleven times greater in volume than all of the land above the sea. But unlike the land, which is populated by life only at its surface, the entire volume of the oceans is a potential habitat.[44]

At the very beginning of our planet's existence, Earth was lifeless and its three organs were in chemical equilibrium. No rocks survive from that distant time 3.9 to 4.65 billion years ago. That's because our restless planet has been continuously recycling itself, so that almost all the physical evidence testifying to the nature of Earth's original crust has been ground to dust, melted and formed anew. But by examining rocks that date to a slightly later time, when Earth's life force was still weak, we can gain deep insights into what the enlivening of our planet meant.

In 2009 I visited the man who pioneered the still controversial idea that life might have helped create the continents. Minik Rosing is the director of the Geological Museum in Copenhagen and one of the foremost authorities on the origin of life. A ponytail- and jeans-wearing Inuit, he's possessed of immense hospitality, and as we sat in his office drinking tea and watching the snow fall outside, he spoke of his love of old rocks. The most venerable surviving parts of Earth's rocky crust are, he said, between 3.3 and 3.8 billion years old. They're precious relics of the youngest Earth we can directly know, formed less than a billion years after the planet itself came into existence. And the very oldest can be found in Greenland.

Minik rose from his seat as he spoke and handed me a rock from his desk. It was, he said, around 3.8 billion years old, and I was astonished to see that it was not folded, battered and scarred as you might expect, but undistorted, its layers as smooth as sheets on a hospital bed. In one layer was a slender black smear, which Minik said marked the start of Earth's carbon cycle, a cycle that largely defines and maintains our planet. Instantly my mind was swallowed by the gulf of time that separates us from that moment when the living Earth-machine first ticked over. Today the carbon cycle runs at full roar, but back then, in a shallow ocean on a planet as fragile as an unshelled egg, it was as delicate and fluttering as a quickening.

Geologists have learned a great deal about the infancy of our Earth through studying such rocks, and no lesson is more marvellous than the strong grip that life has exerted on our planet over its 3.5-billion-year existence. At the time those rocks were formed, and for long after, Earth's atmosphere was toxic, incapable of supporting life as we know it. The oceans also were a toxic brew, with high concentrations of metals such as iron, chromium, copper, lead and zinc, as well as carbon and other elements. All of this changed when microscopic plants and bacteria began to break CO_2 into oxygen and carbon, and to use the metals dissolved in the sea water to speed up the chemical reactions that were essential to their existence. As they died and sank to the ocean floor, they carried their minute cargoes of metals with them, and so, over aeons, the oceans were purged of their dissolved metals, becoming chemically similar to the oceans of today. The metals buried in the sediments had a different fate. Often, they were carried deep into the crust, where heating and compression further concentrated them, leading to the formation of ore deposits. Sometimes these ore bodies became incorporated into the continents and were thrust high into mountain ranges, forming the fabulous golden wealth of places like Telluride, Nevada, or the Incan mines of Peru. A similar process gave rise to Earth's coal, oil and gas deposits, though these formed as a result of living things pulling CO_2 from the atmosphere, rather than from them taking metals into their bodies.

This distant Earth history has profound implications for our modern industrial society. It accounts not only for the state of our atmosphere and oceans, and the good fortune of some countries in possessing valuable mineral deposits, but for our bodies' often-calamitous love of toxic metals as well. All of that matters because today we are digging up these elements at an unprecedented rate, and redistributing them through our air and waters, and that can have surprising consequences. As we will learn later, this is a tale of fundamental planetary disorder, which helps explain why some of us develop disorders such as intellectual disabilities and schizophrenia, and even perhaps why murder rates are high in some communities.

It may seem a paradox that living things should take in toxic metals such as cadmium and lead as avidly as if they were the most precious nutrients on Earth. Assay any one of us and you'll find a treasure trove of toxic metals at concentrations many times greater than they occur in the natural world around us. The answer to the paradox lies in those oceans of long ago. Back then life consisted of little more than bags of chemical reactions floating in an ocean packed with metals. The laws of chemistry dictate that some of the reactions most crucial to life are enhanced by the presence of metals. In technical parlance, metals are catalysts and co-factors—substances that either permit or accelerate chemical reactions. Catalysts are perhaps most familiar to us from the catalytic converters in cars, which work by using a metal, often platinum, to hasten reactions that remove pollutants from the car's exhaust. In our bodies catalysts hasten enzymic reactions and, in an ocean full of potential catalysts, early life became dependent upon them. So unchanging has life's chemistry been over the past two billion years that the majority of the seven hundred-odd chemical reactions that run our bodies today are identical to those that occurred in those bags of chemical reactions that were early life.

As early life mined the ocean's dissolved metals the waters became leached of catalysts and living creatures became desperately hungry for them. Even today, it is metals that limit life's spread in the oceans. In the frigid Southern Ocean, for example, a lack of iron is the key factor limiting plankton growth. Add iron and life flourishes. After two billion years of coping in a world where metals are not easy to be had, life has become extremely adept at keeping hold of whatever metals come its way. And in a world where human activity is releasing metals to the air and seas in ever greater abundance, that can be dangerous, for, like many good things, too much metallic catalyst can be very dangerous. So it is that, despite the damage mercury does to us, our bodies absorb the mercury in the fish we eat even more avidly than the flesh of the fish itself. We store up the metal in our livers, skins and brains, even after we are mortally poisoned by it.

The links between Earth's oceans, crust and atmosphere are nowhere more elegantly exhibited than in the theory of continental drift. Every three hundred million years or so the continents coalesce, creating a single large continent surrounded by oceanic crust. Then the landmass breaks apart again, eventually to come together in another cycle. You can think of the continents acting like dollops of scum floating on a pot of boiling water. The dollops move around, joining together and breaking apart, driven by the convection in the boiling water. While no one understands precisely what drives the movements of the continents, convection within Earth's molten mantle, Earth's gravity, and the pull of the Moon all appear to be factors.

There are two kinds of plates: continental and oceanic. When two continental plates are moving apart, new oceanic crust forms between them. When a continental and oceanic plate collide, however, the oceanic plate is thrust under the continent, and is melted. As a result, mountain ranges, volcanoes and mineral-rich rocks are formed. A good example of this is the Andes. When two continental plates collide, it's far harder for one to slip under the other. Instead the plates buckle, and truly gigantic mountains, such as the Himalayas, are formed. Rivers erode the mountains, creating fresh new soil, and it's this renewal, along with the slow grinding of glaciers, that fertilises life on Earth with the minerals that are essential to plant and animal growth. It's no accident that some of our greatest civilisations sprang up on the plains laid down along rivers flowing from high mountains. If the continents were spawned by life, then we must see this fantastic movement of Earth's plates as at least partly a consequence of life itself.

The most important thing about the movement of the continents in relation to life in the oceans is the effect it has on the recycling of salts. The waters of the ocean are recycled, by evaporation and precipitation and thence through Earth's rivers, every thirty to forty thousand years, and with each recycling rivers leach salt from the continental rocks and carry it into the sea. You might deduce from this that the oceans are growing saltier, and in the nineteenth century this is exactly what scientists thought. They assumed that the oceans contained fresh water upon their formation, and, knowing the rate at which salt is carried into the oceans by rivers, they estimated Earth to be just a few tens of millions of years old. They then coupled this faulty finding with a prediction that a sort of salty doomsday awaited us a few million years hence, when the oceans would become as salty as the Dead Sea.

The truth is far more remarkable. The saltiness of the oceans has remained relatively constant for billions of years, and the drift of the continents plays a vital role in this regulation. As the continents move apart, the ocean's basalt crust is stretched ever more thinly, until it finally ruptures. These rupture lines are known as mid-ocean ridges. They are often located near the centre of ocean basins, and they allow the basins to grow wider. These remote, submarine mountain ranges are rich in life but are among the least known places on Earth. Greg Rouse, a friend of mine, explores them in a submersible, and he's one of the few humans ever to have seen them at first hand.

In 2005 Rouse explored one of the last unknown submarine mountain ranges, deep in the South Pacific Ocean. He showed me video footage taken on the trip of a fantastical white octopus which he had captured using a robotic arm and put in a container on the outside of the submersible. He was wildly excited at the thought of naming and describing the amazing creature, but during the three-hour ascent the ghostly octopus managed to open the lid of its container and escape back to the depths. He also told me something completely surprising. On the evening before one dive, a filleted fish had been cast overboard by a crew member of the support vessel, and when Rouse arrived at the crest of the range four kilometres below, he discovered the filleted fish lying on the bottom, just where the submersible landed. For this to occur, the column of water below the vessel must have been completely serene. To us inhabitants of the turbulent atmosphere such things are utterly astonishing, and they underline how little we understand our planet and its workings.

Mid-ocean ridges form where two continents are moving apart, stetching the oceanic crust between them. They resemble a double-crested mountain range, and between the crests, in a sort of rift valley, molten rock from deep in the Earth's crust comes to the surface. Hydrothermal vents—deep, fluid-filled cracks in the oceanic crust—form, and all of the ocean water in the world eventually circulates through these. It takes between ten million and one hundred million years for all the water to be recycled through the hydrothermal vents, but as it circulates the chemical structure of the sea water is altered by the extreme heat, and the salt is removed. This recycling of the oceans through evaporation, rainfall and rivers every thirty to forty thousand years, and through the crust at the mid-ocean ridges every ten million years or so, keeps the saltiness of the sea constant. And none of it would be possible without continental drift.

What this potted history of Earth tells us is that if we wish to keep our planet fit for life, some of the most routine and humble things we do must change. For as long as we've existed our conception of waste disposal has simply been shifting objectionable matter from one of Earth's organs to another. Whether it's been a human body or a banana skin, we've buried it (returning it to the earth), burned it (returning it to the atmosphere) or tossed it into the sea. On a small scale, this approach to waste disposal works pretty well. But it most decidedly will not do in the twenty-first century, for the very essence of much pollution derives from human actions that weaken the elemental imbalance between Earth's organs. Over the vastness of geological time Gaia's housekeeping has put every element in its place. Carbon has been withdrawn from the atmosphere by plants and geological processes, until just a few parts per ten thousand remain. Iron has been stripped by hungry plankton from the seas, as have mercury, lead, zinc, uranium and a great many other elements, all of which have been safely sequestered deep in Earth's rocks. But now the human burrowers in the Earth have arrived, and, as we tunnel into those buried troves, we undo the work of aeons.

CHAPTER 5

The Commonwealth of Virtue

*So long as a country remains physically
unchanged, the numbers of its animal
population cannot materially increase.*
ALFRED RUSSEL WALLACE 1858

For four and a half billion years our Earth has waltzed around the Sun, and in its silent progress it has given birth to life—at first simple and uncoordinated, but today breathtakingly complex. Life uses Earth's crust as a kind of great rocky shell—a skeleton which it helped to create and to which it is irrevocably and intimately anchored. This shell permits the recycling of elements that are essential to the continued presence of life, and also acts as a vault for locking away toxic compounds. The oceans and other waters are life's bloodstream, helping convey nutrients, heat and elements through the whole, while that most amazing creation of life, the atmosphere, protects, recycles, conveys and clothes. And, ensconced on such a planet, life has ramified and interconnected at every level: from the simple ecology of a bacterial community, to an ant colony, to a bird and so on to the planet itself. But the degree of integration between the numerous and complex parts of these living confederations varies—from loosely organised ecosystems to the intricate integration of an organism such as a human being.

Is Gaia like a human body, an ant colony, an ecosystem—or something else? The physician Lewis Thomas, when investigating this question, wrote of our world, 'We do not have solitary isolated creatures.'[45] Every creature is, in some sense, connected to and dependent upon the rest. 'One way to put it is that the Earth is a loosely formed, spherical organism, with all its working parts linked in symbiosis.'[46] Influenced by Earth's spherical shape, Thomas felt that the closest analogy to Gaia was a single cell. But the most vital words in his lucid description are surely 'loosely formed'. It is the degree of interconnectedness that differentiates ecosystems, insect colonies and organisms. Cells are hardly loosely formed, but it's worth remembering that even the most tightly interconnected organisms evolved from miniature ecosystems. Recall the angophora tree, with its chloroplasts that were once bacteria. The cells of our bodies provide another excellent example. They are composed of two or more entirely separate and unrelated types of creatures which must once have been relatively independent parts of an ecosystem on a young Earth. The bacterial partner is known as mitochondria, and it's what gives cells energy. These partners must have started by forming a loose association, but after more than a billion years of evolution they have become the indivisible parts of an organism, or rather every organism on the planet. Every multicellular being is descended from such a union.

There are clues in today's living world of how this remarkable partnership may have started. Corals are polyps that resemble miniature sea anemones. By themselves they are colourless. All the colours we see in coral reefs come from tiny algae that live inside the coral polyps' cells. These algae benefit the coral by helping feed it, and they in turn gain shelter. While the relationship is intricate, it is not indissoluble—the coral polyp can temporarily expel the algae if it suits it. Left without algae for too long, however, the coral polyp starves and coral bleaching is the result. The difference between this relationship and that between the cells of our body is that the once free-living bacteria that are our mitochondria have been with our cells for so long that neither can survive—even for an instant—without the other.

The complexity of relationships doesn't stop there. Larger living thing are themselves composites, involving whole eco-systems of bacteria, fungi and invertebrates. Without many of these creatures—our gut bacteria, for example—we could not exist. These fellow travellers make up 10 per cent of our weight, and are so pervasively distributed over our bodies that were we to take away all 'human' cells, a detailed body-shadow composed of them would remain: we are in ourselves virtual planets of Gaian complexity.

Time is important in forging such interdependence, and Gaia is very old—a quarter as old as time itself. But has Gaia, over that near-eternity, established the kind of interdependence characteristic of an organism? Profound organisational rules guide the develop-ment of cooperation, and as our bodies have grown more complex the specialised cells that form our brain and nerves have created command-and-control systems. Such systems, even if they are not created by nerves and brains, are the hallmark of the organism, being present in all but the most rudimentary of multicellular animals.

Entities such as ant colonies lack command-and-control systems, but they do have means of coordinating their activities. It's astonishing to consider that the intricate workings of an ant colony, with its millions of individuals and hundreds of metres of inter-connected tunnels, is maintained without command and control. But such is the case, for there is no brain caste or blueprint for the colony resident in any individual ant brain. Instead, the activities of the ants are regulated via chemicals, known as pheromones, which they create and disperse. These chemicals elicit a specific response from an ant, so if one comes across a particular pheromone, say on an ant trail, it will behave in a certain way. While rudimentary by comparison with the functioning of a nervous system, these chemicals allow sufficient coordination for colonies of millions of individuals to function.

Gaia clearly lacks a command-and-control system, but perhaps it possesses something roughly equivalent to pheromones. These are substances, created by life, which can act to maintain conditions on Earth favourable to life. Among the most important are ozone, which shields life from ultraviolet radiation, and the greenhouse gases CO_2, methane and nitrous oxide, which play a critical role in controlling Earth's surface temperature. Dimethyl sulphide is another. It is produced by certain types of algae, and it assists in cloud formation. Clouds play a vital role in the Earth system by bringing rainfall, shading vulnerable organisms and altering Earth's brightness (its albedo). Atmospheric dust, much of which is organic in origin, or derived from rocks through weathering processes, may also have a regulatory effect.

Clearly these substances differ from pheromones in that they do not elicit responses from other individuals, but from the Earth system as a whole, and they are often multiplicitous in their impact. But they do have a pheromone-like effect, in that they contribute to the maintenance and good running of the entity. And like pheromones they are potent—even minute traces can generate a powerful response. For these reasons I think it appropriate to refer to these substances as 'geo-pheromones'.

One of the foundation stones of Lovelock's Gaia hypothesis is the idea that Earth regulates its surface temperature to favour life.

But how well does it do it? Self-regulation (or homeostasis) gives some indication of how integrated an entity is. Warm-blooded (homeothermic) organisms such as ourselves maintain a constant body temperature regardless of external conditions. And some of the most complex ant and termite colonies manage to do the same thing, the colony's internal temperature being controlled by a sophisticated architecture that rivals that of skyscrapers. Is Gaia similarly competent?

Lovelock's first indication that Earth could regulate its surface temperature came from the 'faint young Sun' paradox, which states that although the Sun was 25 per cent cooler three billion years ago than it is today, the average surface temperature of the Earth has stayed within a narrow range. This seems a convincing argument for efficient self-regulation, but on shorter time scales Earth's average surface temperature varies enormously, the shift from ice age to interglacial conditions (such as we are now living in) being a fine example. Every hundred thousand years for the past million years, Earth has experienced a cycle of gradual freezing followed by abrupt thawing, which involves a rapid increase in average temperature of 5° Celsius—from 9° to 14° Celsius. This huge increase occurs within around five thousand years, and is driven by small changes in Earth's orbit, its tilt on its axis and its 'wobble' on its axis which occur in cycles known as the Milankovitch cycles, after their discoverer the Serbian engineer Milutin Milankovitch. Their impact on the amount of energy Earth receives from the Sun is small—just 0.1 per cent—though where the sunlight falls varies more. The Arctic receives more summer sunlight during that part of the cycle where Earth's tilt is greatest and the Arctic is nearest the Sun. But such large changes resulting from minute variations suggest, initially at least, that Gaia's ability to control its temperature is inferior to that of the most highly evolved superorganisms.

James Lovelock, however, has a different explanation. He points out that Gaia is composed of two parts, each with different requirements, and that the huge temperature shifts which occur as glacial periods gives way to interglacials result from one of these entities gaining supremacy over the other. The two entities Lovelock refers to are life in the sea and life on land. Life on land prefers an average temperature of around 23° Celsius, for that's the temperature at which land-based plants flourish. Life in the oceans, in contrast, prefers a more chilly 10° Celsius or less: at such temperatures the ocean's surface and bottom waters can mix through convection, bringing nutrients to the surface. If Lovelock is correct, then our Earth has a two-state thermostat, which results from a colossal tug of war between life on land and in the sea, each pulling Earth's temperature towards its preferred state, the balance of power being altered by the tiny variations brought about by the Milankovitch cycles. This might not seem to reflect the condition of most organisms and superorganisms such as ant colonies. But consider the reptiles, or even hibernating mammals. They also exist in one of two states, warm or cold, depending upon external conditions.

The founder of sociobiology, Oxford University's Bill Hamilton, was one of the greatest biologists who ever lived. After a lifetime studying colony-forming insects such as ants and bees, in his fifties he turned his mind to the nature of Gaia, and he did so in an intriguing way—by considering the role that life might have in contributing to atmospheric convection. With his co-author Tim Lenton, he wrote that:

> Herrings falling with rain miles inland in Scotland, frogs
> and a juvenile turtle being found in American hailstones,
> and live bacteria and fungal spores collected by rocket more

than 50 km from the Earth's surface all demonstrate that
both terrestrial and marine organisms are sometimes raised
very high by extreme atmospheric events.[47]

But what causes such vigorous circulation? Hamilton and
Lenton concluded that plankton and bacteria, emerging from the
ocean's surface as bubbles burst, contribute to cloud formation by
becoming nuclei for water droplets. In effect they had discovered
some of Earth's more important geo-pheromones. But they also
believed that they'd discovered the reason that these microscopic
organisms have evolved properties that enable them to behave this
way. In brief, Hamilton and Lenton argued that Earth's convection
is assisted by plankton and bacteria because a vigorous convection
helps them disperse, thus giving them a truly global reach. We
can imagine the bacteria and plankton being carried high into the
atmosphere, then falling over Earth's entire surface reaching every
suitable habitat available to them. The concept neatly synthesises
Darwin's and Wallace's thoughts on dust, though neither man is
referred to in Hamilton and Lenton's paper. But Hamilton and
Lenton go further than Darwin or Wallace, speculating that the
various kinds of plankton might work as teams, some species
helping to produce winds, and others ice nuclei, which drop the tiny
travellers back to the surface. In summarising their work, they said:

> The mechanisms we describe do not directly bring us any
> nearer to discovering why life influences that are stabilis-
> ing to the planet should be more common than destabilising
> ones...But a proof that large side effects, stabilising or not,
> can arise from activities that are adaptive...for thoughtless
> aerial and marine plankton, strengthening the expectation
> of large influences from similar unpromising systems, can
> perhaps help clear a path towards a principled theory.[48]

That principled theory, which might have explained Gaia, may well have been discovered by Hamilton had his research not taken a fatal turn. Shortly before his death he wrote to Lenton that he was excited about a computer program that seemed to show that as ecosystems become more complex they also become more stable and productive. As he put it:

> A Genghis Khan species may be less likely to be about to destroy life on the planet than I had previously speculated... Even with an 'unexpandable resource base' to the model we do find some accumulation of resistance to disaster from the next species added; and when the model is endowed from the first with physical possibility that its resource base can be expanded if the right species are acquired, very disturbance-resistant communities sometimes appear.[49]

Such stability, of course, is the very essence of Gaia theory, and to find it in a computer model more complex than Daisyworld (which did not impress Hamilton) seemed a major step forwards.

Then, in the 1990s, Hamilton became obsessed with an idea worthy of Wallace himself—that the origin of HIV lay in oral vaccines given to children in the 1950s—and it led him to Africa in search of proof. While in the Democratic Republic of Congo he contracted malaria and was evacuated to Britain. Just six weeks later, on 7 March 2000, he died of cerebral haemorrhage. Ever the entomologist, he left a sum in his will for:

> My body to be carried to Brazil...It will be laid out in a manner secure against the possums and the vultures...and this great *Coprophanaeus* beetle will bury me. They will enter, will bury, will live on my flesh; and in the shape of their children and mine, I will escape death. No worm for me nor sordid fly, I will buzz in the dusk like a huge bumble

bee. I will be many, buzz even as a swarm of motorbikes, be
borne, body by flying body out into the Brazilian wilderness
beneath the stars, lofted under those beautiful and un-fused
elytra which we will all hold over our backs. So finally I too
will shine like a violet ground beetle under a stone.[50]

Had Hamilton lived to combine the 'narrow roads of gene
land', as he characterised his genetic research, with the expansive
vision of Wallace and Lovelock, he may have become the most
revered biologist of all time.

In his absence we continue to struggle with the question of
whether Gaia is akin, in its level of organisation, to an organism, an
ant colony or an ecosystem. It seems to me that the level of organisation that can be achieved using geo-pheromones is perhaps best
described as a 'commonwealth of virtue'. In such a commonwealth
the various elements are sorted and stored in the most appropriate
planetary organ. Non-living parts of the system are coopted for
the benefit of life, and there is no 'waste' because species recycle
the by-products of others. And there is a tendency, over time,
towards increased productivity and interdependence. All of this
is achieved in the absence of a command-and-control system, and
with only limited ability to elicit specific, system-wide responses.
The remaining question, as Hamilton realised, and which we shall
re-visit towards the end of this book, is whether a commonwealth
of virtue so defined promotes its own stability: in other words, is it
Medean or Gaian in nature?

A commonwealth of virtue as I've defined it is also pretty much
what an ecosystem is, although the complexity of ecosystems varies
enormously. So how are ecosystems formed, and what binds them
into coherent entities? Many of Alfred Russel Wallace's seemingly
off-the-cuff observations led to deep thought. In the brief essay in

which he introduced his ideas on evolution, he speculated on the nature of evolutionary processes in a country that remains physically unchanged.[51] It's an intriguing idea—just letting heavens' performance run on and on, without the interruptions and extinctions brought about by a restless Earth or colliding asteroids.

When we think of the evolutionary process, we most often imagine organisms evolving in response to changing conditions. Our own human lineage provides a good example, its evolutionary history having been shaped by a drying climate in East Africa. But a drying climate is just one of many possible physical changes that can shape the evolutionary process. Movements in the Earth's crust or changes in sea level can separate islands from continents, so segregating populations of animals and plants, and exposing them to different environmental pressures. Flightless birds such as the dodo, which were once found on islands round the world, are just one example of a response to such changes. Left without predators on their island homes, those individuals that put more effort into reproducing, and less into flying, are successful.

And of course the reverse can occur. Land bridges can open, allowing a mixing of species that had evolved separately for many millions of years. Such changes usually result in extinctions, as superior predators and competitors displace less well adapted ones. Extinctions also result when large asteroids strike the Earth, and here there are no immediate winners, but, as the destruction of the dinosaurs shows, species evolve to take advantage of the ecological niches vacated by the asteroid's victims. None of these events, however, would occur in a country long unchanged, which doesn't mean that evolution stops—only that different pressures drive the process. And over time they forge the intricate relationships that lie at the heart of Gaia.

It's axiomatic that the selective forces operating in a country long unchanged come from other organisms—there is no other source of challenge to drive selection. From viruses to potential mates, other living things have an impact upon the reproductive potential of an individual, and so drive the process of evolution by natural selection. Among the most important of these pressures is that exerted by members of the opposite sex. Known as sexual selection, it's a subject that fascinated Darwin. Sexual selection results from members of one sex finding certain individuals of the opposite sex more attractive than others. The peacock offers a fine example. Females are attracted to males with colourful plumage and 'eye patterns' on their tail feathers. Males with the longest, most colourful and 'eye-filled' tails leave the most offspring, resulting in today's peacocks, with their spectacular colours and cumbersome tails.

As is evident from this example, sexual selection can result in individuals that bear handicaps (the male peacock's tail hinders it as it moves about). But unless counteracting selective pressures, such as predation, are sufficiently strong, sexual selection will continue its work, often producing seemingly disadvantageous traits.

Humans offer an interesting example of sexual selection. In many traditional societies men go to considerable lengths to control the sexual and therefore the reproductive potential of women, including enforcing celibacy until marriage, dire penalties for adultery and even surgical procedures such as clitoridectomy. While such attempts can never have been entirely successful, there's no doubt that historically they have limited the mate choices available to women. Within the last few decades in western societies, however, women have, by and large, gained control of their reproduction. Liberated and armed with contraceptives, they now represent a powerful evolutionary force that is busy shaping the men of tomorrow. That's because, through the men they choose to father their children, women are manifesting in flesh the ideal mate (or as close as they can attain to it) that exists in their minds. Over evolutionary time this selection must and will change the nature of men.

Another driver of evolution in a country long unchanged results simply from the different rates at which species evolve. The last survivors of many evolutionary lineages are giants—the horses, rhinoceros and apes (including ourselves) are good examples. In the past, these lineages consisted of both small and large species, so why are they now limited to just a few giants? Small organisms reproduce more rapidly than large ones, which allows selective pressures to be exerted on a larger number of generations over a period of time. Thus, all else being equal, smaller organisms evolve faster than large ones. Where the ecological niches of smaller and larger organisms overlap, this advantage allows the smaller organisms to displace their larger, slower breeding competitors. Over evolutionary time they increase their own body sizes, and expand their ecological niche to overlap with ever larger competitors. Eventually, they replace all but the very largest members of the slower-breeding lineage. In the case of the horses and rhinos it was the cud-chewing grazers such as cattle and sheep that displaced their smaller relatives, while for the apes it was the Old World monkeys.

Natural selection that is triggered by interactions between related things is called coevolution. It can act at every level, from that of individual amino acids to entire organisms, and it may not be just a property of life, but something far more profound. Astronomers argue that black holes and galaxies develop an interdependence that's akin to biological coevolution. Indeed Erich Jantsch, in his book *The Self-Organizing Universe*, attributes the development of the cosmos to coevolutionary forces.[52]

Coevolution was what Bill Hamilton was investigating using computer models just prior to his death. In the real world, it can lead to the development of ever more intricate relationships, which can, in some circumstances, create a sum of biological productivity that is greater than its parts. Take, for example, the microrrhizal fungi that sheath the rootlets of the scribbly gum. Similar fungi partner with many kinds of plants that grow in poor soils, and together, even where soils are appallingly infertile, fungi and plants can create spectacular biodiversity. Indeed, the biodiversity of Africa's fynbos and Western Australia's heathlands rivals that of the rainforests. Likewise, it's the partnership between the coral polyp and its algal partner that creates the diversity of a coral reef. At a more humble level, pastoralists have always known that you can feed more cattle per hectare by sowing the seeds of half-a-dozen species of grasses rather than just one. Importantly for our future, coevolution tells us that nature's bounty is not inflexible, but is instead a kind of magic pudding that can be made to expand if cooperation between species is fostered.

In a country physically long unchanged, coevolution can produce complex ecosystems that seem to have reached an equilibrium, having altered very little in their overall structure for long periods. This is precisely what we see in the tropical rainforests of places like South-East Asia. They're filled with species, the Sumatran rhino among them, whose near relatives are found in European fossil deposits tens of millions of years old. Europe has changed greatly in that time, but South-East Asian rainforests hardly at all. It's not surprising, therefore, that it's in the rainforests, with their immensely long histories, that we find the most intricate examples of coevolution, such as the orchids and their insect pollinators.

These relationships can be extraordinarily specific. Some orchids, for example, fool wasps into attempting to mate with the stamen, and so carry pollen on specific parts of their bodies to other flowers of the same species. This phenomenon so fascinated Darwin that he wrote a monograph about it, which included a remarkable deduction.[53] He was aware of a brilliant white orchid flower from the island of Madagascar, which had protruding from it a long, spine-like nectary. Knowing that some insect would have to reach the nectar at the bottom of this structure to reap its reward for fertilising the flower, Darwin deduced that a moth with a proboscis twenty-five centimetres long existed on the island. Forty years later, long after the great man had died, the moth was discovered.

Coevolution can also lead to a kind of arms race, in which species adapt to advances made by others. On the African savannah, lions catch only the old or weak. Antelopes in their prime keep just ahead of the lions. After all, if the lions could catch those in their prime with little effort, their prey would become extinct, while if the antelopes invested energy in running far faster than lions, they'd be wasting effort. It was in such a coevolutionary world that our lineage spent at least seven million years—that's how long there have been upright apes in Africa. And all the while we've been coevolving with other African creatures—predators, prey and diseases. Coevolution explains why Africa alone among the continents retains its full diversity of large mammals. They have got to know us as predators, and as part of an evolutionary arms race they've evolved means to avoid us, which is very different from what has happened on other continents.

An astonishing example of coevolution developed in our African homeland. The greater African honeyguide, an undistinguished-looking bird of medium size, feeds solely on the larvae, wax and honey of beehives. It often attacks the hives in the evening, when a drop in temperature makes the bees lethargic, or after some larger creature such as a honey badger has damaged a hive, but when a honeyguide encounters a human, it sees an opportunity. Uttering a striking call, which it otherwise uses only in aggressive encounters with other honeyguides, it attracts the human's attention, then moves off, stopping frequently to ensure that the person is following it, all the while fanning its tail to display white spots that we visually oriented humans find easy to see. When native Africans reach a hive with the help of a honeyguide, they break it open and often thank the bird with a gift of honey.

After hundreds of thousands of years this unique relationship is beginning to break down, for in many areas sugar is becoming cheap and widespread, and people are less willing to exert themselves in the pursuit of honey. Faced with lazy humans, it seems, the honeyguide is giving up on its coevolved partners in the hunt.

I believe that coevolution, in both a biological and a cultural sense, is critical to our hopes for sustainability. Indeed, I think that our environmental problems ultimately stem from having escaped coevolution's grip, for we humans have a gypsy history, and as we've spread across the globe we've broken free of environmental constraint and destroyed many coevolutionary bonds that lie at the heart of productive ecosystems.

2

A TURBULENT YOUTH

CHAPTER 6

Man the Disrupter

We live in a zoologically impoverished world,
from which all of the hugest, fiercest, and
strangest forms have recently disappeared.
ALFRED RUSSEL WALLACE 1876

What is it, precisely, that makes us humans different? Is our cultural mode of evolution the magic ingredient, as it allows us to evolve far faster than organisms reliant on mere evolution by natural selection? Yet this cannot be the whole story. All higher species have cultures, and their cultures too change and evolve in response to the environment. If you doubt this, consider the deer. Mature white-tailed deer stags have magnificent ten-point antlers that are prized by human hunters, and the deer have developed a cultural means of avoiding a dose of lead in the heart. This was discovered during an experiment in which stags were penned with human hunters in a 2.5-square-kilometre enclosure, surrounded by lookout towers. Observers were surprised to find that many stags survived by lying completely still, so that human hunters stalked right past them.[54] Such a strategy would mean instant death for a stag pursued by wolves or cougars—predators that hunt by smell. But somehow stags have learned, then presumably taught each other, that this is the best way to avoid being killed by visually oriented humans.

This, however, falls far short of humanity's evolutionary strategy. What we have done is combine cultural evolution with technology in a way that allows us to mimic key aspects of evolution by natural selection, and speed it up ten thousand times. Thus we make spears rather than evolve fangs, and weave clothes rather than grow furry coats; and it's this ability, which has been with us from the very beginning, that makes humans so formidable.

Earth's history is a story of coevolution, punctuated by disruptions caused by asteroids, abrupt climatic shifts or invasive species that break apart the threads that lie at the heart of ecosystem function. Such disruptions invariably result in an impoverished world, in the short term, at least, because they lower productivity and drive species to extinction. Of all the disruptions Earth has suffered, few can compare with those we have caused, for by virtue of our discovery of a new way to evolve we have become a destructive force par excellence. Over the past fifty thousand years our Medean tendencies have repeatedly been unleashed, leaving in their wake a world of ecological wounds.

Evidence of humanity's power to disrupt ecosystems dates almost to the moment our species left its ancestral African homeland. We've eaten our way through one resource after the other as we've spread around the planet, and only after long experience in one place have we acquired the wisdom of managing the land. As a result, it is our misfortune to be only now, perhaps, tentatively emerging from a world in which human genius was so without wisdom that it fractured and disfigured nature's evolutionary bonds to the point of our own self-destruction.

Our kind arose in Africa, as savannah-living apes, and it's clear from the rubbish dumps left behind that even 1.8 million years ago our ancestors were capable hunters and ardent carnivores. By around two hundred thousand years ago, when the first humans had evolved, we had become such expert killers that creatures much larger than ourselves, including the largest of all land mammals, were forming part of our diet. From an ecosystem perspective, this ability to hunt the largest and fiercest creatures was destined to become, and until ten thousand years ago to remain, our defining ability. Yet when our species left its natal continent of Africa our hunting skills began to tear apart ecosystems, and the further we wandered, the more god-like our control over the fiercest and largest of creatures became.

Genetic studies allow us to read in detail the migrations of our ancestors. The Y-chromosome is passed down solely through fathers, making it an ideal guide to the travels of our male ancestors. Fortuitously it evolves quickly—around 40 per cent of its mutations have occurred since humans began to exhibit regional variation ('race' in the old parlance). Studies of the Y-chromosome reveal that all people alive today are descended from a single male who lived in Africa around sixty thousand years ago.[55] This genetic Adam, it's reasonable to deduce, was dark-skinned, tall, slender, and possessed epicanthi, the folds over the corner of the eyes commonly seen in people from Asia.

But what of Eve? The mitochondria can help us here. There are none in the sperm head, so fathers don't contribute to their offspring's mitochondria, which makes it ideal for tracing female ancestry. The tale told by mitochondria confirms our African origin. But there is one startling difference between our male and female ancestry: mitochondrial DNA tells us that all people alive today can trace their ancestry to an African woman who lived at least a hundred and fifty thousand years ago. Adam and Eve, it seems, never met, being separated by ninety thousand years.[56] The explanation for this lies in the fact that small populations tend to lose genetic lineages faster than large ones. We might expect men and women to have existed in roughly equal numbers in the past, but it is the size of the breeding population that really counts. The population of breeding men, it seems, has long been small relative to that of women, probably because an exclusive group of high-status men has tended to father most children.

Genetic studies reveal that Africa has repeatedly acted as a fountainhead of hominid dispersal. The first successful diaspora of humans occurred around fifty thousand years ago, when a few clans left Africa and began to spread east. A particular genetic marker on the Y-chromosome, known as M130, originated in a man living somewhere north of the Red Sea in the early days of this migration, and this marker is now widely distributed across southern and eastern Asia, as well as in Australia. The American geneticist Spencer Wells postulates that the people bearing this mutation were coastal dwellers adapted to harvesting marine resources, an ecological niche that would have facilitated a rapid spread eastwards.[57]

The peopling of inland Eurasia, and eventually the Americas, was accomplished by a separate group, whose ancestor was an African man who bore a Y-chromosome marker known as M89. This dispersal involved people who became adapted to drier, inland conditions. When they reached the Middle East they gave rise to a further mutation, known as M9, and then to three further separate Y-chromosome mutations, the bearers of which went on to settle central Asia, Europe and India. Y-chromosome studies also reveal that it's only over the last ten thousand years that humans from outside Africa were able to migrate back into their ancestral homeland, presumably because they had developed agriculture and animal husbandry.

It's ironic that just as we've found a way to read the genetic chronicles documenting our ancestors' travels, increased human mobility is leading to such a mixing of our genes that, within a few dozen generations, the clues to our ancestors' wanderings will be obliterated. Then, we will have returned to the genetic state that prevailed before we ever journeyed out of Africa: all humans forming a single, genetically uniform population.[58]

I sometimes wonder what the world would have been like if it had taken our ancestors ten times, or even a hundred times, as long to evolve the technology required to colonise the world. What would it have been like if Columbus, for example, had sailed to the Americas one million years, rather than around thirteen thousand years, after the first Americans had arrived from Eurasia via the ice-age land bridge? A million years is more than enough time for evolution by natural selection to have forged Americans and Europeans into separate species, and if that had happened a smile or a gesture might not have been understood, and sex would not have been a common currency. Had prehistory not unfolded as it did, our global civilisation might not have been possible.

Wherever our ancestors arrived, they fundamentally altered the environment. And as a result, the world we live in today is sadly truncated—all of its largest, fiercest and most striking creatures are no more. Gone are the moa of New Zealand, the mastodons and sabre-tooths of the Americas, the giant marsupials of Australia and the gorilla-sized lemurs of Madagascar. Only in our species' homeland of Africa does anything like the beastiary known to our distant ancestors survive. Until recently it was widely disputed whether humanity could have caused such changes. But the evidence is now overwhelming. Indeed, it's not just humans who, in the right circumstances, can bring about such changes, but even creatures as humble as toads.

The cane toad was introduced into Australia via Hawaii in 1935. Until then, no member of the toad family (Bufonidae) had existed in Australia, meaning that no Australian predator had any experience of the venom that toads produce from the glands on their necks. Originally inhabitants of Central America, the toads were introduced to help eradicate pests in sugarcane, a job they failed dismally at. But they spread quickly, causing environmental destruction wherever they went. In northern Australia it's easy to know when you are ahead of the toads. You see crocodiles, goannas, frill-necked lizards, birds of prey and countless other native Australian creatures every day, all around you. But if you visit the region behind the moving frontier (which in 2010 lay near the Western Australian border), all is silence and devastation. Unless you were there when the first toads arrived, it's hard to fathom what happened. A friend of mine was camping on a river in western Queensland when the toads went through. He went out one evening to fish, and was distracted by a nauseating smell. Following it upstream he discovered a logjam of dead crocodiles, their bodies lying so thickly that they clogged the stream. The toads had arrived, and so sensitive were the crocs to the toads' poison that a single toad was enough to kill even the largest of them.

Because cane toads metamorphose from the tadpole stage while very small, around the size of your little fingernail, they represent a danger even to the tiniest of predators, which accounts for the eerie emptiness of the lands behind the toad frontier. A visitor to the toad's homeland of Central America would never suspect that cane toads could be so calamitous to so many creatures, for in their ancestral homelands they are relatively rare and live in harmony with the myriad alligators, lizards and other species that share their environment. Coevolution over millions of years has bestowed resistance to toad poison on many. Either that, or they have learned not to eat toads. In Australia no predator knew instinctively that the toad was a threat—a naïvety that caused their demise.

The toad pioneers, those at the very front of the invasion wave, have longer legs than the settler toads that populate the land behind, and they are more aggressive. On the frontier, evolution by natural selection favours toads with the drive and physique to invade quickly and take advantage of the bounty of resources awaiting them.

For humans on the frontier, it's not our legs that respond to evolutionary opportunities, but our culture and our attitude to life. In frontier societies, individuals who can monopolise the greatest bounty, regardless of waste or environmental consequences, are selected as 'fittest'. And for most of human history the bounty our ancestors sought was big, hairy and dangerous. As our ancestors spread out of Africa, they encountered creatures with no experience of modern humans. But the animals they met did know about upright apes, for a distant relative of ours had left its African homeland nearly two million years earlier. *Homo erectus* had settled the more temperate portions of Europe and Asia, extending as far north as modern-day Beijing and southern England, and as far south as Java.

I have great difficulty in deciding how to address *Homo erectus*. What would I say if I met one in the street? Is it an animal, or a man? I use 'it' here only because I wish to emphasise the 'us' in this story of humanity's spread. Part of my difficulty comes from the fact that *Homo erectus* changed over its 1.8-million-year existence, developing several regional types, one of which, that inhabiting Africa, gave rise to us. But it's the *Homo erectus* population of Europe and Asia that is important to this story. *Homo erectus* hunted the large mammals of Eurasia for a million or more years, thus providing everything from mammoth to cave-bear with experience of carnivorous apes, and starting an evolutionary arms race that would allow some species to avoid extinction when our kind arrived.

With a brain around 25 per cent smaller than our own, Eurasian *Homo erectus* doubtless lacked many advantages possessed by our ancestors. One of the best documented concerns stone tools. Those made by *Homo erectus* are abundant, providing a detailed record of change over time. They slowly improve in quality, and by three hundred thousand years ago *Homo erectus* was making a greater range of tools than was being produced 1.5 million years earlier. But even at its best, *Homo erectus'* tool kit was rudimentary when compared to that of our ancestors. A similar picture emerges with regard to fire, which was haphazard or absent 1.8 million years ago. There is evidence (albeit still disputed) that by three hundred thousand years ago *Homo erectus* was using fire to some extent. But its control of this vital tool was rudimentary compared to that of our ancestors.

In other ways, however, *Homo erectus* never approached our achievements. Studies of the skull and neck suggest that it was unable to speak, though it may have communicated in other ways, such as sign language, that have left no fossil record. And of course *Homo erectus* left us no art, or evidence of care for the dead.

Much is often made of these differences. But what is really important from an environmental perspective is the almost identical ecological niche occupied by *Homo erectus* and our own ancestors. Like us, *Homo erectus* was a hunting-and-gathering ape, and from the beginning it was capable, in some circumstances at least, of hunting prey as large as young elephants. From the perspective of the large hairy beasts of Eurasia's forests and plains, *Homo erectus* was simply a less capable version of our ancestors. If you want to imagine what those beasts experienced as *Homo erectus* gave way to *Homo sapiens*, imagine playing in the amateur league at whatever sport you prefer, then being suddenly matched against the professional elite in a game where not just egos but life itself is at stake. At least you'd be better off than those who had never played the game at all, which was the position of the megafauna of the Americas and Australia, where no carnivorous ape had ventured prior to our arrival.

There were probably animal victims of *Homo erectus'* expansion into Eurasia. The sabre-tooth cats vanished from the Old World at around the time *Homo erectus* arrived, yet they survived in the Americas until our species showed up thirteen thousand years ago. Some of these sabre-tooths specialised in killing young elephants, which would have brought them into direct competition with *Homo erectus*. A few potential prey species also became extinct at around the time of their spread, but the evidence is so subtle and it occurred so long ago that it's hard to say whether the extinction resulted from *Homo erectus* or other factors such as a changing climate. What we do know, however, is that by the time our ancestors arrived in Eurasia around fifty thousand years ago, *Homo erectus* had struck a balance with the surviving large mammals, for by then they had coexisted with Eurasia's elephants, rhinos and other great beasts for almost two million years.

CHAPTER 7

New Worlds

O wonder!
How many goodly creatures are there here!
How beauteous mankind is! O brave new world
That has such people in't!
SHAKESPEARE *THE TEMPEST*

One of the great surprises of archaeology is the discovery that our ancestors colonised Australia a full fifteen thousand years before they settled western Europe. The possessors of Y-chromosome variation M130 had moved swiftly from the Sinai to Sulawesi, and just a few thousand years after leaving Africa were poised to invade the Great South Land.[59] I can imagine them scanning the seemingly limitless ocean to the south looking for signs of another virgin land to settle. The forebears of the people who stood on that beach had already crossed the sea time and again—from Borneo to Sulawesi perhaps, or Bali to Lombok, and then onwards across strait after strait, each time discovering islands full of food naïve to the human hunter.

Imagine leaving the first footprint ever on a tropical beach. The sand is strewn with the seashells of the largest size, and lying in the shallows is seafood for the taking, all in quantities beyond reckoning. Fish cruising the shallows would not flee at the sound of a spear being hurled, but would instead gather round the victim. We know this to be true from eighteenth-century accounts of Europeans who landed on coral atolls never before visited by humans. They wrote of sharks seizing their oars as they paddled ashore, of tame birds that sat on their nests until plucked off like ripe fruit, and of such an abundance of turtles, fish and shellfish as to glut the entire crew.

The first humans to leave Africa would have learned this quickly. In their journey from Suez they must have pillaged many a near-shore reef and atoll. And as they progressed east they must have faced ever longer and more perilous sea voyages. But reward outweighed risk, and they pushed on. And, every time, the arrival of the upright apes was a disaster for the land they discovered. Hardly a creature in the new-found lands had an adequate defence against them, and there was nothing to stop our ancestors consuming all they required, then moving on. And so this virulent manifestation of *Homo sapiens* (a creature we might call *Homo medeaensis*) was born on the frontier—a frontier that was to roll on, in one form or another, into the twenty-first century.

The sea barrier separating Australia from island Asia was more formidable than anything *Homo sapiens* had crossed before. For millions of years a veritable menagerie of capable swimmers, such as elephants, had existed on the Asian side of the ocean, but none except those famous island colonists the rats, which drifted there on rafts of vegetation, had reached Australasian shores. The one thing those first colonising humans had which other creatures lacked was a culturally reinforced will to colonise. And so, between forty-five and fifty thousand years ago, they set out. What a journey it must have been, so ambitious that no one had ever attempted anything like it—or at least not survived if they had. The voyagers would have lost sight of their homeland some time before they were able to make out the land whose existence they had deduced lay ahead. With nothing around but boundless sea, the journey must then have entered a time of crisis. From the very beginning it was a leap of faith, and now they were engulfed in the maw of the ocean. Perhaps a smudge of dust or a plume of smoke sustained hope and saw them press on into the emptiness. Such a perilous voyage would surely have been undertaken only by the greatest of risk-takers— and their reward would be the inheritance of an entire continent.

The story of how those voyagers changed the land they discovered is only now beginning to emerge. Back then Australia was home to an enormous array of giant marsupials, birds and reptiles, including marsupial equivalents of rhinoceros, hippopotamus, giant sloths, leopards and antelopes. They shared the land with giant emu-like birds, huge goannas, horned turtles and gigantic, primitive snakes, as well as the distinctive Australian species of today, such as kangaroos and koalas. Back then Australia was a *Through the Looking-Glass* version of Africa's Masai Mara. It had been isolated from the rest of the world for fifty million years, and none of its creatures had ever seen an upright ape or anything like it, for *Homo erectus* had never ventured so far.

By forty-five thousand years ago, within a few thousand years of the arrival of humans, a rapid and dramatic extinction had stripped the continent of this marvellous diversity of megafauna, leaving nothing larger than a human or a red kangaroo in the land. In all, about sixty species of giants vanished, along with an as yet uncounted number of lesser creatures. They disappeared so swiftly that some scientists have referred to their loss as a blitzkrieg extinction, an event that happened within a few hundred years.[60]

The consequences were to be profound. Australia is a fragile land of infertile soils and variable rainfall, and in such circumstances ecological misfortunes tend to amplify. By eating vegetation that would otherwise burn, the marsupial giants had suppressed fire. They recycled the few nutrients available in the vegetation through their guts, pouring out fertiliser in the form of manure and urine, thereby increasing the carbon content of soils, enhancing soil moisture and promoting plant growth.

With their extinction, however, the grasses and bushes grew rank and fires raged, baring soils, causing erosion and destroying nutritious but fire-sensitive plants. Humans had, in effect, taken the ecological bankers out of the ecosystem, and overall productivity most probably diminished ten- or a hundredfold. With its complex and sustaining ecosystems in tatters, Australia became a land of fire, whose soils grew ever more impoverished, and whose once-living carbon had joined a flux of fire-generated CO_2 pouring into the atmosphere.

At 7.9 million square kilometres Australia is the smallest of the continents, and it's unlikely that this transformation of vegetation into CO_2 had a global impact. But it may have influenced Australia's climate. Prior to the arrival of humans, northern Australia was covered by a kind of rainforest that may have shed its leaves in the dry winter. Much like the forests of the Amazon, which create 80 per cent of the rainfall over the Amazon basin today, these ancient Australian forests enhanced rainfall by transpiring moisture. But when they burned, their capacity to enhance rainfall was lost, and so the great lakes that lie at the continent's heart dried up.

In the Northern Hemisphere, by thirty-seven thousand years ago humans had reached what is now Romania on the eastern frontier of Europe, but there is no unequivocal evidence of them having reached western Europe earlier than thirty-two thousand years ago. Shortly thereafter, however, they rapidly spread, displacing the Neanderthals that had until then held Europe as their own. The Neanderthals represent yet another dispersal out of Africa by an upright ape. Genetic evidence indicates that they were close relatives of ours, having split from our lineage just half a million years ago, somewhere in Africa. They settled a swathe of Eurasia north of the Himalayas all the way into Europe, in the process presumably displacing some populations of *Homo erectus*, but also settling in climates too frigid for those earlier invaders.

The ecological niche of the Neanderthals is fascinating. They took to an extreme the big-game hunting that our lineage excels at. Some scientists, observing that the skeletons of male Neanderthals are as full of broken bones as those of a veteran rugby front-row forward, think that they leaped on mammoths and stabbed them to death. Whatever the case, their hunting was not sufficient to drive Europe's big game to extinction, for mammoths, straight-tusked woodland elephants and two species of woodland rhinoceros coexisted with Neanderthals for hundreds of thousands of years.

The fate of the Neanderthals remains deeply mysterious. Their final stronghold was the rocky slopes of Gibraltar, where a few bands clung on until twenty-five thousand years ago. Did our ancestors kill them all, or interbreed with them, so absorbing them into our gene pool? Our genes are 99.5 per cent the same as theirs, so interbreeding was certainly possible. Researchers have located portions of the genes of Europeans that they believe are extremely ancient, perhaps bits of Neanderthals hiding within us. But as yet the evidence is too incomplete to be conclusive.

The fate of Europe's big game is clearer. The straight-tusked elephants become extinct on the mainland of Europe around thirty thousand years ago, and by about twenty thousand years ago the two European species of rhinoceros had followed. After that, in the European region, elephants and other slow, large creatures survived only on uninhabited islands such as Crete, Sardinia and the Dodecanese Islands. One by one these islands were settled and their dwarf elephants hunted to extinction. By four thousand years ago, they survived only on the island of Tilos in the Dodecanese. The dwarf elephant of Tilos was a pony-sized creature, related to Europe's straight-tusked elephant, and it might be with us today if humans had overlooked its island home for a few thousand years longer.

During the ice age a dry, frigid grassland known as the mammoth steppe stretched east and north from forested Europe. The largest single habitat on land, it extended from what is now central France to Alaska, and so harsh were conditions there that it resisted colonisation by all bipedal apes—our species included—until after thirty thousand years ago. It is true that various hardy bands may have attempted summer sallies onto it to reap a rich harvest of big mammals, but the lack of wood and shelter made this region too hostile for permanent colonisation. And so the last great herds of Eurasian mega-mammals lived on. Woolly mammoth, woolly rhino, bison, musk ox, giant elk and horse abounded there in a climate far harsher than today's.

Siberia is now incapable of supporting such mammals. The soils are too acid and the nutritional value of the boggy plant matter is too low. The only herds remaining are reindeer (known as caribou in North America), which eat lichens rather than grass. It's difficult to imagine how grass-eating woolly mammoths, woolly rhinos and bison could have abounded in such a place, yet their bones regularly emerge from the permafrost to remind us that this was once a fantastically productive landscape.

While woolly mammoths were large, they're hardly giants of the elephant family, being similar in size to their living Asian elephant cousins. Yet they differed in a number of striking ways. Depictions of them still loom on cave walls as if out of a snowstorm, their humped shoulders, long, shaggy fur and extraordinary curved tusks making them look like ghosts adrift on an ocean of snow. Their metre-long hair coat was greasy, and, like musk oxen, they probably shed the underfur in spring. Their crossed tusks, which still lure ivory traders to their frozen graves, were great snow ploughs they used to expose the grass in winter, and they left feeding opportunities in their wake for horses and bison unable to dig so deep themselves. The large hump on the mammoth's shoulders was fat, a food store for surviving the winter and doubt-less relished by Neanderthal and human hunters alike. There are also features you wouldn't notice on a cave painting. Under the mammoth's tail was a plate-sized stopper that fitted snugly over the anus, a heat-saving device that was lifted only when nature called. And the tip of the trunk was not pointed, like those of living elephants, but wide and flattened to assist in gathering grass.

How could the harsh, northern Siberian ice-age environment have supported such stupendous creatures? Scientists call this the productivity paradox—because it's productivity rather than cold that limits animal populations—and they have identified three factors that may have made the mammoth steppe more productive than today's tundra.[61]

The first relates to the fate of plants after they die. Dead tundra vegetation doesn't get a chance to rot; instead, it becomes water-logged and freezes, eventually forming peat. The frozen layer thickens and never defrosts, locking away nutrients and freezing the soil beneath. The peat is also highly acidic, making it a hostile environment for plants even in the brief summer.

The second factor concerns the functional length of the growing season. Today's climate is warmer and wetter than when the mammoth steppe existed, but the growing season for plants may be shorter than it was in the past. That's because when summer came to the mammoth steppe, the Sun could reach the bare soil quickly, warming roots and releasing nutrients from the soil. Today the frozen peat delays the warming of the roots, so reducing the growing season.

The third and perhaps most decisive factor is that there were more available nutrients for plants back then. Today, frozen peat and high acidity make the tundra an infertile place in which plants struggle for phosphorus and nitrogen. On the mammoth steppe, however, a copious flow of urine and mammoth droppings provided large volumes of these nutrients at the soil surface. The mammoth, in other words, was the engine of productivity on the mammoth steppe. If a creature ever deserved the epithet of biological banker, it was the mammoth. By eating the vegetation that would have become peat, and returning it as fertiliser, it kept a climatically formidable environment productive and alive.[62]

By the coldest phase of the ice age, around twenty thousand years ago, the woolly mammoths and woolly rhinos had already been hunted out of the more favourable southern parts of their range and survived only in the harshest environments of the far north and east. But our ancestors were tinkering away with pelts and stone, discovering ever more effective ways to clothe themselves against the cold, and devising more efficient weapons and strategies for turning mammoths into fat and meat. Then, around fifteen thousand years ago, the climate began to warm, and by fourteen thousand years ago humans, moving ever eastwards, had pushed as far as the eastern, ice-bound cul-de-sac of the mammoth steppe that is today Alaska. Back then it was connected by an ice-free plain to what is now Siberia, but to the east and south it was bounded by walls of ice. When they arrived, the humans destroyed the herds, leaving only the truly forbidding lands of Yakutia and the Gydansky and Taimyr peninsulas, which jut into a frozen Arctic sea, as a last refuge for the steppe's mega-mammals. It was here, between 11,000 and 9600 years ago, that the last of the mainland mammoths made their final stand.[63]

As with Europe's dwarf elephants, some island populations of mammoths managed to hold on after all mainland populations had become extinct. One survived until around six thousand years ago on St Paul's Island in Alaska's remote Pribilof chain. Another inhabited a remote island high in a frozen ocean at the top of the world known as Wrangel. There, mammoths lived in splendid isolation until at least 3700 years ago. These were the very last of Earth's dwarf island elephants, and they survived for so long that the Pharaohs, had they been so inclined, and venturesome enough, might have entertained themselves with the sight of hairy elephants the size of circus ponies. But theirs was the last generation of humans to be afforded such an opportunity. It was only a matter of time before skin-clad, spear-carrying hunters trekked over the winter ice to Wrangel and discovered this last bounty.

As the mammoth steppe was being torn limb from ecological limb some fourteen thousand years ago, there remained one last region of the planet innocent of human impact. South of the ice barriers in what is now Alaska lay two entire continents—the Americas—and they were filled with creatures that had long vanished elsewhere. Mammoths, mastodons (a primitive member of the elephant family) and sabre-tooth cats abounded, as did American novelties such as giant sloths, short-faced bears and dire wolves. Indeed the ice-age fauna of North America was as varied and spectacular as that of the Serengeti. But by 13,200 years ago the ice barrier that separated this naïve world from humans was melting away.

The first North American pioneers developed a distinctive culture known as Clovis, which flourished for just three hundred years. Evidence of it is found right across North America in the form of its signature piece: lethal-looking stone points, some of which have been found wedged between the ribs of fossilised mammoths, making their purpose evident. The impact of Clovis' hunting upon the fauna seems to have been devastating and immediate. The horses, camels, sloths and other giants vanished rapidly. It took a little longer—until around 12,800 years ago—for the last of the mammoths and mastodons to become extinct, and the continent's most ferocious predator, the giant short-faced bear, may have survived in the deserts of the southwest for a thousand years after that. Because the classification of the large, extinct mammals is still incomplete, it's hard to know precisely how many species were lost, but within five hundred years of human arrival North America had lost thirty-four genera of large mammals (each genus being composed of one or several related species) weighing more than forty-four kilograms, while South America lost fifty genera, the most of any continent.[64]

Bones are a haphazard way of understanding the past. We'd be extraordinarily lucky to discover the skeleton of the last mammoth to live in North America. But there are other ways of determining when the large beasts vanished. *Sporormiella* is a fungus that grows on animal dung, and where the dung piles are great the fungus is abundant. *Sporormiella* spores abound in the dung piles of Shasta ground sloths preserved in caves in Utah, and they're well represented in the sediments of North American lakes that accumulated before 12,900 years ago. But after that the spores virtually disappear, only to reappear with the advent of cattle. In Madagascar, *Sporormiella* tells the same story. It declines within a few centuries of human arrival, and reappears with the introduction of cattle about a thousand years later. The decline of *Sporormiella* tells us that North America's great herds had been cut down, right across the continent, in a geological instant.[65] And yet a few species did survive, and they are as informative of the cause of the major extinction as is the Clovis point.

Among the survivors were ground sloths, monkeys and giant rodents related to guinea pigs, all of which lived on the islands of the West Indies. They thrived for four thousand years after every ground sloth, mammoth and other large mammals between Alaska and Tierra del Fuego had vanished. Then, around eight thousand years ago, humans discovered the Caribbean Islands and the beasts vanished into the same black hole that had consumed their mainland relatives—the hole located between nose and chin in a human being. Humans ate them all, down to the last one.

If more evidence is required of the nature of this extinction, it's provided by the bison, brown bears and moose. Today these are the largest creatures native to North America, yet all are new-comers, having arrived from Eurasia over the very same land bridge that brought people into the New World. Indeed the bears and the moose most likely arrived at the same time as the people, for there is no fossil evidence of their presence in the Americas prior to this. The reason that they survived, when so many indigenous species perished, is that they had coevolved with *Homo erectus* in Eurasia, and so had learned how to avoid upright apes.

A curious codicil to this story concerns horses and camels, both of which originated in North America before a few lineages invaded Eurasia prior to the arrival of *Homo erectus*. These kinds of creatures were always far more diverse in the Americas. The fact that only the few lineages that made it to Eurasia (including the horses, donkeys and camels) survived, and were then able to thrive in North America when re-introduced, is also due to their long period of coevolution with *Homo erectus*. It's striking that very similar species that thrived in North America up until thirteen thousand years ago didn't stand a chance.

In Eurasia there was no blitzkrieg extinction, but instead a slow grinding away of many species, as one after another lost their arms race with a culturally evolving humanity. And some of the shaggy monsters did not give in easily. The giant unicorn was a bizarre kind of rhinoceros that once roamed central Eurasia. Fossils of it are rare (though the Natural History Museum in London has a splendid skull), making it one of the least known of Eurasia's megafauna. But it was also one of the most spectacular. At two metres high, six metres long, and weighing up to five tonnes (a similar weight to an elephant), it was shaped unlike any other rhinoceros, its long legs giving it a horse-like gait. But its most remarkable feature was doubtless the two-metre-long horn that adorned its head.

In the tenth century an Iraqi traveller recorded what may have been the last remnant population of this spectacular beast. In the summer of 921 Ibn Fadlan departed Baghdad as part of an embassy to the king of the Bolgars. Two years later he reached the Volga River north of Samara Bend, and there he heard tell of a fabulous creature, of which he wrote:

> There is nearby a wide steppe, and there dwells, it is told, an animal smaller than a camel, but taller than a bull. Its head is the head of a ram, and its tail is a bull's tail. Its body is that of a mule and its hooves are like those of a bull. In the middle of its head it has a horn, thick and round, and as the horn goes higher, it narrows (to an end), until it is like a spearhead. Some of these horns grow to three or five ells [an ell is the length of a man's arm], depending on the size of the animal. It thrives on the leaves of trees, which are excellent

greenery. Whenever it sees a rider, it approaches and if the rider has a fast horse, the horse tries to escape by running fast, and if the beast overtakes them, it picks the rider out of the saddle with its horn, and tosses him in the air, and meets him with the point of the horn, and continues doing so until the rider dies. But it will not harm or hurt the horse in any way or manner.

The locals seek it in the steppe and in the forest until they can kill it. It is done so: they climb the tall trees between which the animal passes. It requires several bowmen with poisoned arrows; and when the beast is in between them, they shoot and wound it unto its death. And indeed I have seen three big bowls shaped like Yemen seashells [giant clam shells], that the king has, and he told me that they are made out of that animal's horn.[66]

Perhaps stories of this amazing beast helped fuel the myth of the unicorn. Whatever the case, Ibn Fadlan's account is the only written evidence we have of its possible survival into modern times.

But what of the fate of *Homo erectus*? At one time it was thought that the various populations of *Homo erectus* might have evolved separately into *Homo sapiens*. Under this hypothesis, Peking man (from caves near Beijing in China) was considered directly related to the modern Chinese. Genetic analysis has conclusively disproved the idea. But still, we know almost nothing of the fate of populations such as Peking man's. We can only surmise that, like the Neanderthals, *Homo erectus* gave way as *Homo sapiens* spread.

The very last of the non-human bipedal apes to escape extinction as our ancestors spread around the globe inhabited the island of Flores. Part of an island chain that stretches eastwards from Java towards New Guinea, Flores and its neighbouring island of

Komodo are perhaps most famous as the home of the komodo dragon. Until 2003 it was assumed that Flores was one of a series of stepping-stone islands that humans used to reach Australia. But then, a team of Indonesian and Australian archaeologists digging in a cave discovered the skeleton of a small upright ape that dated to seventeen thousand years ago. Named *Homo floresiensis* by the scientists, it has become more widely known as the hobbit.[67]

Hobbits were diminutive creatures, weighing just nineteen kilograms and standing at about the height of a three-year-old child. While humanoid in appearance, their arms were longer than ours, and their wrists chimp-like. Yet it's the hobbit's brain, as revealed from casts of the inside of a complete skull, that astonishes most. It was tiny, around a quarter the volume of our own, but with frontal lobes—the seat of planning and rational thought—uniquely specialised and enlarged. Were hobbits rational and intelligent in an entirely different way from us? We can only speculate.

The hobbit's ancestors left Africa around two million years ago, earlier even than *Homo erectus*, and somehow they managed to cross the sea from Java and Bali to Lombok, then to Sumbawa and Flores. Volcanic eruptions periodically exterminate life on Lombok and Sumbawa. But Flores is a less explosive isle, and there for a million years or more the ancestors of the hobbit lived. They adapted, as mammals on islands often do, by shrinking in size—perhaps in response to a limited food supply.

When hobbits first arrived on Flores, the island was home to a straight-tusked, pygmy elephant and a giant tortoise, rather like that which can still be seen on the Galápagos Islands. Both must have swum to Flores, and both were hunted to extinction by ancestral hobbits. Later, a second species of straight-tusked elephant was able to colonise the island, and this species, which presumably had experience of *Homo erectus* on Java (and was larger than the pygmy species) was able to survive hunting by hobbits. The diminutive hominids nevertheless took their toll. Judging from the debris they left in caves they specialised in killing newborn elephants, and also hunted the cat-sized rats and komodo dragons that abounded on the island.

Flores' komodo dragon and giant rats have long posed a zoogeographic riddle. We know from fossils that both were once widespread throughout the island chain stretching east of Java—the komodo dragon (or a very close relative) even inhabiting Australia—until the first humans arrived.[68] Why should such creatures survive nowhere but Flores? The answer comes, I think, from the hobbit. For a million years the dragon and the giant rats on Flores experienced being hunted by an upright ape that was almost certainly not as competent as our ancestors. This experience was a sort of primary school for the larger animals. The giant tortoise and original Floresian pygmy elephant failed to survive, becoming extinct more than eight hundred thousand years ago, while the larger pygmy elephant that arrived later did survive the hobbits. But it did not learn quickly enough, and died out when humans arrived on Flores. The dragon and the large rats, however, learned well enough to survive the human invasion. Their existence is thus testimony to a vanished island world that was shaped by our diminutive hominid cousin.

It seems likely that the hobbit survived in Flores until around twelve thousand years ago. By then, our ancestors had long been firmly established in South-East Asia, New Guinea and Australia. Flores and its hobbits thus were surrounded by humankind. But for some reason ancient humans did not invade their island realm. Some think that fierce ocean currents kept our kind away, but the island's isolation remains an enduring mystery, as does the nature of what happened at the moment of reunion between hominid lineages that had been separated for two million years. We can only imagine what went through the minds of the first humans to walk a Flores beach. What did they make of this land of pygmy elephants, giant lizards, and tiny humanoids armed with spears and other tools? Was curiosity, fear, loathing or hunger uppermost in their minds?

Perhaps one day the caves and swamps of Flores will yield their bounty, allowing us to know how the hobbits perished. At present all we can say is that the hobbits have gone and that we, in glorious, self-created isolation, rule the world as the only surviving upright ape.

Though gone, the hobbits may have left us a living legacy. We humans are vexed by two species of lice, which went their separate evolutionary ways around two million years ago. Genetic studies reveal that one of them is a long-established hitchhiker on human skin; but the other is a newcomer, having first climbed aboard our kind towards the end of the ice age. A surprising number of human parasites occur as pairs of closely related species: tapeworms, follicle mites, dysentery-causing organisms and bedbugs as well as the lice. Could one of each species pair, all those years ago on Flores, have jumped from a dying to a thriving race? Whatever the case, they should remind us that every event over our long history has had its legacy.[69]

As I reflect on this sorry tale of environmental destruction, it occurs to me that some of the changes brought about by our ancestors were potentially large enough to have affected Earth's climate. After all, the mammoth steppe was the largest single land-based habitat on the planet, and it was (seasonally at least) very productive. Could its destruction have altered Earth's carbon balance? Looking back at the record of climatic change revealed in the ice cores retrieved from polar icecaps, it turns out that there is some evidence that this might be the case. The cores record the history of Earth's climate going back at least 740 thousand years, and within them we see the glacial cycles repeating over and over, bringing an ice age followed by a sharp warming every hundred thousand years thanks to the Milankovitch cycles, which conspire to deprive the Northern Hemisphere of sunlight in summer. This means that not all of the snow that fell the previous winter melts, allowing the icecaps to grow until the entire planet is in their thrall. But then, abruptly, Earth warms, and over the next hundred thousand years the pattern is repeated.

The amount of CO_2 in the atmosphere varies with the Milankovitch cycles. During ice ages there are fewer than two hundred parts per million because lots of carbon is transported into the ocean depths. This happens because cold water can hold more CO_2 than warm water (think of opening a warm can of soda, and how rapidly the gas escapes) and because oceans that are cool all the way to the surface have convection currents that can carry the CO_2-rich surface water to the bottom. But then, as Earth warms, CO_2 levels rise to between 280 and 300 parts per million.

Not every Milankovitch cycle produces identical changes in Earth's climate and atmosphere, though the broad pattern is the same. But in the most recent cycle, which ended with the abrupt warming of ten to fifteen thousand years ago, atmospheric CO_2 peaked at only 265 parts per million. That's at least fifteen parts per million short of where it would usually have been, making for a somewhat cooler interglacial period. So what happened to the missing carbon? One possibility is that some of it was trapped, in the form of peat, across the region that had been the mammoth steppe. But detailed studies are lacking, so it's not yet clear whether the amount of carbon in Siberian peat bogs is sufficient to account for the discrepancy. Nevertheless, it's an intriguing possibility that humanity, even as hunter-gatherers, may have been capable of influencing Earth's climate.

In Earth's atmosphere we can read the relative health of our living planet. In this first example, a deficiency of fifteen parts per million of CO_2 may tell of the extirpation of the mammoth. Whatever the cause, it was a deficit of particular importance. As James Hansen argued in his book *Storms of My Grandchildren*, the cool start to the interglacial period allowed sea levels to stabilise and remain stable for eight thousand years.[70] And, as we shall soon see, this in turn permitted the growth of human superorganisms.

CHAPTER 8

Biophilia

A new commandment I give unto you,
That ye love one another.
JOHN 13, KING JAMES BIBLE

Since our emergence from Africa fifty thousand years ago, the sweep of human history has been a tale of destruction that has crippled one ecosystem after another. So what justifies the view that humans have the capacity for a cooperative, sustainable, Gaian future? Those who killed the last mammoth, or the last hobbit, had no idea that they were extinguishing a species, much less altering ecosystems, for their world view extended only as far as their clan boundaries. To concentrate only on the destruction they wrought is a bit like understanding evolution's cruel mechanism without acknowledging its legacy. As the human frontier passed, those people left in its wake found themselves reliant on an impoverished ecosystem from which extinctions and dislocations had removed many resources. It would have been possible for those hunter-gatherers to continue their lethal course. But instead, in most instances, coevolution slowly drew them into a balance with the new ecosystems they had helped create.

The problems faced by hunter-gatherers trying to manage their resources are immense, for they do not control them as a farmer does. After all, a bison or kangaroo can wander from one clan territory to another, and he who kills it first benefits. The problem of how humans have managed commons (and many of the resources of hunter-gatherers are types of commons) has been the lifetime study of Elinor Ostrom, the 2009 Nobel Laureate in economics. She discovered that sometimes humans do manage commons sustainably and has elucidated the conditions under which we are most likely to succeed. In essence, she says that there is no golden rule, but that certain conditions favour commons' successful management, including the ability to exclude outsiders; clear, mutually agreed rules about who is entitled to do what, along with appropriate penalties for transgressors; an ability to monitor the resource; and mechanisms to resolve conflict.[71] Some of these conditions exist for certain hunter-gatherer societies, and, in order to understand how they contribute to the maintenance of ecosystems, we need to visit humans as they existed in Wallace's hypothetical land long unchanged. In Australia and the mountains of New Guinea, until very recently, the oldest human cultures on Earth have altered little for tens of thousands of years.

The Australian Aborigines are widely known for managing their land using fire. The practice, known as firestick farming, consists of a careful and precise system of burning which is aimed at (among other things) producing useable vegetable resources at the required times and the maintenance of marsupials important as food. In the first half of the nineteenth century the Australian explorer Sir Thomas Mitchell saw the system in action, and had this to say of it:

Fire, grass, kangaroos, and human inhabitants, seem all dependent on each other for existence in Australia; for any one of these being wanting, the others could no longer continue…the native applies that fire to the grass at certain seasons, in order that a young green crop may subsequently spring up, and so attract and enable him to kill or take the kangaroo with nets. In summer, the burning of the long grass also discloses vermin, birds' nests etc., on which the females and children, who chiefly burn the grass, feed.[72]

Through their use of fire, Aborigines had become ecological bankers in the Australian environment, for fire consumed the vegetation that the megafauna would otherwise have eaten, and so recycled the nutrients it held. This was instrumental in maintaining Australia's biodiversity. As the term firestick farming suggests, the Aboriginal use of fire resembled agriculture in some ways: it yielded certain crops at certain times, suppressed weeds and was carefully controlled. But did Aboriginal people have the kind of control that Ostrom suggests can lead to well-managed commons? Aboriginal people are fiercely protective of their clan lands, excluding outsiders or inviting them in as conditions warrant. There are also clear rules about who has the right to what resources and highly evolved mechanisms to resolve conflicts and enforce penalties. This has enabled Australia's Aboriginal people to act as keystone species of the continent's ecosystems for forty-five thousand years. As the Europeans displaced them, Australia's fragile environment collapsed into a far less productive and diverse state.

There is much more, however, to the management of biodiversity in these lands long unchanged, and some of the most striking examples concern taboos whose observance requires a considerable sacrifice on the part of individuals, yet provides no direct individual benefit. The reason I'm interested in these practices is threefold: firstly they are informative of how evolution by natural selection can result in cooperation; secondly they have resulted in the preservation of megafaunal species of great importance to the ecosystems they are part of; and thirdly they are informative of the way that cultural and physical evolution, working together, may provide a means of humans nurturing the ecosystems that support them.

The Telefol people inhabit the very centre of the island of New Guinea. Like Australia's Aborigines, they practise initiation rituals that bestow increasing status on older men. The most senior have passed through six such initiations, all of which take place in a cult house that the Telefol believe to be the navel of the Universe. In order to get to the village and the cult house, men must pass through a sacred grove of native pine trees that is protected by the strictest taboo. I have walked through the grove with Telefol and noticed that they do not disturb even a single leaf. Several kinds of birds of paradise inhabit the grove, and the gaudiest males gather there, displaying their finery with complete impunity.

Other Telefol taboos protect vulnerable creatures far from the village. One of the most intriguing involves the long-beaked echidna. A relative of the platypus, this large egg-laying mammal is covered in spines and snuffles about in the mossy forest feeding on worms, which it locates with its sense of smell and possibly an electromagnetic sensor in its snout. Once found in Australia and New Guinea, it became extinct everywhere except New Guinea's mountains over forty thousand years ago. Weighing up to sixteen kilograms and almost a metre long, it is a slow-breeding, defence-less creature that can live up to fifty years. And because its flesh is the fattiest and tastiest in all of New Guinea, even modest human populations are sufficient to completely exterminate it. Yet not so long ago it abounded close to villages in the land of the Telefol, for Telefol hunters refused to harm it.

When I questioned people about their reluctance to hunt long-beaked echidnas, they told me a story about their ancestress Afek. Afek had four children: an opossum-like marsupial known as the ground cuscus, a rat, a human and a long-beaked echidna. The long-beaked echidna was his mother's golden boy. She loved him best of all, but the smoke from the fire irritated his weak eyes, so she told him sorrowfully that he must leave the family home to live in the high mossy forest, where the air was cleaner. As the long-beaked echidna departed Afek warned her remaining children that they must never, under any circumstances, injure their brother, for if they did disaster would befall all Telefol. Until the 1950s Telefol hunters found it unthinkable to harm a long-beaked echidna. But then a Baptist missionary arrived in the region, and the Telefol were taught that their ancestors' beliefs were the work of the devil. Within a short space of time long-beaked echidnas simply ceased to exist in the region.

It's worth emphasising how easy it would have been for a Telefol hunter in pre-missionary times to kill a long-beaked echidna without anyone knowing. Telefol men often hunt alone, and they can camp for days in the high country, consuming meat as they go, before returning to their home base with smoked rations for the family. No one would know if they had eaten echidna. So why did hunters not cheat? The reason, I think, is the power of belief. Telefol had no doubt that to kill a long-beaked echidna would call down disaster upon the entire clan. The previous abundance of echidnas indicates that this belief was so strong people rarely, if ever, tested it.

Traditions that protect various animal species were extra-ordinarily widespread in Australasia. In the Popondetta area of southeast New Guinea, for example, people refused to harm sea turtles—a fine source of food—in the belief that they were malign spirits that could wreak revenge. In New Guinea and Queensland entire mountain summits were taboo, providing refuges for species that had become endangered elsewhere, while across Australasia local totem animals, from bats to birds to fish and large game, were strictly protected. As extinctions in the wake of colonisation all too often show, many such species are extremely vulnerable to hunting. Moreover, many were keystone species whose extinction caused a cascade of ecosystem changes that reduced biodiversity and productivity, which must ultimately affect the human societies that depend upon those ecosystems.

To protect such important species, people make real sacrifices, for animal protein is almost always scarce in such societies. Yet the survival of the likes of the long-beaked echidna indicates that people have been doing so for thousands, if not tens of thousands, of years. Such practices seem entirely contrary to any notion of survival of the fittest or our destructive Medean natures. So how might they have originated?

One striking aspect of the societies that share such practices is that they show respect to clan elders by reserving special food for their exclusive use. This is especially true of the Telefol, and generally the larger and rarer the creature, the more restricted is its consumption. I was once in a mountain camp with a dozen or so Telefol when a tree kangaroo was caught. Tree kangaroos are the largest marsupials found in the Telefol lands, and they are extremely prestigious catches. Its meat was duly cooked, and when a plate of it was served to me I was astonished to see that no one else was eating it. A man explained to me that there was no Telefol in the camp who was sufficiently senior to partake; indeed, there were only two men in the entire region who qualified to eat that kind of food, and both resided in a distant village. The rest of the meat would be carried to them, they said. But I, as a non-Telefol, was exempt from the taboo. That I as an outsider was not bound by the taboo indicates that its significance lies solely in relationships between Telefol. It also illustrates how fragile such beliefs are. In the case of the long-beaked echidna, some newly baptised Telefol, who were still squeamish about eating the animals, would nevertheless capture them alive and sell them to outsiders living at the mission station.

Some years ago I discussed this puzzle with Jared Diamond, who asked whether the flesh of the long-beaked echidna might not be so prestigious that, since some Telefol golden age perhaps, nobody has possessed sufficient status to eat it? After all, the Telefol would not be alone in believing in a golden age when leaders were more noble and wise than anyone alive today. The idea makes a lot of sense. In all the instances I know about, it is the largest (and thus most prestigious) species that are reserved for the most senior people and are thus protected. So it may be that the very most prestigious species have simply migrated off the menu.

There are parallels between these practices and nature conservation in Europe and China. One of the most important means of preserving wildlife in these regions has been the royal hunting reserve, which is little more than a food store and recreation area reserved for the most respected members of society. People in today's democratic societies might shudder at the thought of such institutions—they bring to mind social inequality and the image of a nobleman enjoying his sport while the masses starved. It's worth remembering, however, that as power devolved from the nobles to the people, royal hunting reserves often became national parks or green spaces enjoyed by all of society. Indeed central London's great public parks were once royal hunting reserves.

Of all the royal hunting reserves of Europe, the most significant from the perspective of biodiversity conservation was doubtless the Bialowieza forest on Poland's border with Belarus. Famous as the last refuge of the European wisent (a relative of the American buffalo), it illustrates more piquantly than any other reserve the vital role that such places have played. Two thousand years ago the wisent was the most abundant of all the large mammals of Eurasia, with herds spread from Britain to eastern Siberia. But by around a thousand years ago wisent had vanished from western Europe, and, despite the theoretical protection they received in eastern Europe as property of the nobility, their decline was relentless.

The wisent of the Bialowieza forest had been protected by royal decree since 1538, but the breakdown in law and order that accompanied World War I rendered such protection useless. The German army seized the area in 1915, and within months about two hundred wisent had been killed for food. Despite the issuing of an order protecting them the slaughter continued, and just one month before the Polish army recaptured the area, in February 1919, the last wild wisent was killed and eaten. Following this catastrophe, there were no wild-living wisent on Earth. Today there are over three thousand wisent in reserves and zoos—all descended from only twelve individuals that had been brought into captivity. In 1951 zoo-bred wisent were reintroduced to Bialowieza and with human assistance they are slowly recolonising their ancestral realm. Free-ranging herds now roam reserves from Slovakia to Russia.

China has its own story of megafaunal survival in royal hunting reserves. Père David's deer has specialised hooves that allow it to move swiftly through swamps. An inhabitant of the fertile and densely populated coastal plains, the last wild-living individual was killed near the Yellow Sea in 1939, and there the species might have ended had it not been for a Chinese emperor. Many years earlier he had established a population in the imperial hunting park (Nan Hai-tsu) near Beijing. This large park was surrounded by a 72-kilometre-long fence, and entry by the public was strictly forbidden.

On 17 May 1865 the French missionary Père David convinced guards to allow him a peek over the wall. To his astonishment he saw a herd of previously unknown deer. Several pairs were then secured as diplomatic gifts and sent to Europe. By 1900, with the destruction of the fence by flooding and political upheaval, the deer in the park had all died. Just a few dozen descendants of the original diplomatic gifts survived in Europe, and zoo directors decided to send all eighteen fertile animals to the Duke of Bedford's park, Woburn Abbey, in England. Just one stag and five hinds eventually bred, and a mere fifty of their descendants survived World War I. In 1956 four individuals were returned to Beijing Zoo, and in 1986 another twenty-two were returned and released in the same imperial park that had sheltered the species for so long. Subsequently, a wild herd has been established in the region where the last wild animal was seen.

From such practices has sprung a globe-straddling system of parks and reserves that are vital to the protection of biodiversity. Indeed almost every large, wild-living mammal on Earth today owes its survival to the human impulse to preserve rather than destroy, making it arguable that the kind of self-control practised for so long by New Guinean hunters is all that prevents their loss. Another reason I believe that such practices are relevant to us is that they teach us that respect for nature and respect for our fellow human beings are inextricably interwoven. In today's world, social disruption—in particular, conflict flowing from poverty and inequality of opportunity—is the greatest threat to the survival of the last of the megafauna, such as the rhinoceros and elephant.

But those ancient practices just might teach us something more—that people blessed with healthy, diverse ecosystems are likely to endure and prosper. I say this because environments with intact keystone species are more productive, and therefore better habitats for humans. When human groups come into conflict, those from superior habitats, who are better fed and sheltered, are likely to prevail. Not enough is known of the history of New Guinean societies to test this idea in a conclusive manner, but it's interesting that the Telefol are a highly successful people—having colonised adjacent valleys through warfare in pre-colonial times—and that they continue as a powerful regional influence today.

Many practices of traditional cultures act to strengthen eco-system function. In North America, for example, hunters would often spare a few beavers from each lodge. I think that as the ances-tors of the native Americans, Europeans and others first settled into their newly won lands all those thousands of years ago, similar practices sprang up everywhere, and all too often it was only when a new wave of the human frontier rolled through those lands that those practices ceased. Then, humans would have to learn all over again the lessons of living with the land.

Our deep urges to preserve biodiversity prompted the evolution-ary biologist E. O. Wilson to postulate that each and every one of us possesses a fundamental love of nature that was forged during our species' distant evolutionary past. He called this love 'biophilia', and he outlined his hypothesis in 1984 in a book of the same name.[73] Whenever we appreciate nature, protect it or even help an individual wild or domestic creature, Wilson sees biophilia at work. Richard Dawkins, however, is sceptical of the concept, questioning whether evolution by natural selection could ever produce such an outcome.

Whatever its merits in explaining our relationship with nature, the biophilia hypothesis as Wilson outlines it provides a power-ful evolutionary explanation of some of our preferences. Take our desire for a home with a green lawn in front of it and an elevated view of water. This, Wilson explains, reflects the preferences of our African ancestors, to whom short green grass meant the presence of large grazing mammals and a camp with a view of a waterhole that they were ideally placed to hunt the creatures as they came to drink. Those humans who could occupy and defend such locations thrived, and their preferences were inherited by future generations.

The term biophilia was not coined by Wilson, but by the German social psychologist Erich Fromm. In an addendum, known as The Humanist Credo, to his book *The Heart of Man* he said of the concept:

> I believe that the man choosing progress can find a new unity through the development of all his human forces, which are produced in three orientations. These can be presented separately or together: biophilia, love for humanity and nature, and independence and freedom.[74]

Fromm's biophilia beckons us much further than a mere evolutionary explanation—towards a new unity, one with another, and of all with the Gaian whole. Perhaps we are destined to be either Medean or Gaian and never anything in between, and if so Fromm's biophilia describes our only pathway to survival, which is towards a future in which human and environmental health are inextricably linked. But it also reinforces the importance of cultural belief. I find it intriguing that Telefol protection of biodiversity, even if it has an ultimately Darwinian explanation, was executed via a belief—in the ancestress Afek.

3

EVER SINCE
AGRICULTURE

CHAPTER 9

Superorganisms

Ants, like human beings, can create
civilisations without the use of reason.
B. HÖLLDOBLER AND E. O. WILSON 2009

We must now turn to a critical development in the evolution of life—the rise of the superorganism. This momentous event took place long before the arrival of our own species on the scene. Sometime between 120 and 190 million years ago, some cockroaches began to live in colonies and feed on rotting wood. Various members of the colony then became specialised, performing certain tasks, such as defending, foraging or reproducing, for the group as a whole. Eventually some colonies domesticated certain types of fungi, and began to grow extensive fungi gardens, upon which the colony fed. The cockroaches had evolved into termites, and the world's first superorganism had been born.

Termites and other social insects differ from us in many ways, but, like us, they are parts of superorganisms, and they remain an excellent guide to understanding the forces that shape our societies. Yet language sometimes gets in the way. We speak of humans having 'civilisations' and insects as being mere 'superorganisms', for example. This is just hubris: whether human or insect, the greater social entities thus created involve relationships based on identical underlying principles.

The study of superorganisms began in the early twentieth century. Eugene Nielen Marais was an Afrikaans doctor and lawyer who for many years lived in a lonely hut high in South Africa's Waterberg. He befriended a troop of baboons that so grew to trust him that he could walk among them and handle them with impunity. But his great interest was his study of the termites that abounded in the region. His extraordinary book *The Soul of the White Ant* explains that termite colonies are much like human bodies, their workers and soldiers functioning like our blood cells, and their winged kings and queens just like our sperm and eggs. But he also believed that so complete was their oneness—their integration—that termitaria had evolved a psyche that was 'far beyond the reach of our senses'. Marais made one major error in his work—he believed that the termite queen acted as the colony's brain. But the error gives Marais' account a touching reverence. When, after hundreds of attempts, he finally penetrated the royal chamber without disturbing the workers, he wrote as if he had removed a piece of skull, describing the massive, pulsating termite queen as a living, working brain.[75]

The concept of the superorganism is central to our understanding of interconnectedness. It is emblematic of the coevolutionary capacity to create an entity that is more competent and productive than the sum of its parts. Superorganisms comprise individuals whose degree of integration and organisation sits between that of an ecosystem and a multi-cellular creature such as ourselves. Some kinds of organisms are more likely to transform into superorganisms than others. Ants, wasps and bees belong to an insect group called the Hymenoptera, which has given rise to almost all of the insect superorganisms (the exception being the termites). Only 2 per cent of insect species have developed superorganisms, but they have become extraordinarily successful, making up over 30 per cent of the entire animal biomass (the combined weight of all animals) in some Brazilian rainforests.[76]

Like humans, ants started out as hunter-gatherers. Indeed, the most successful and diverse of all the ant groups, the ponerines, continue to live in small hunter-gatherer bands of a few tens to a few thousand individuals. Like the stone-age human hunters who specialised in killing woolly mammoths, most ponerines hunt only one or a few kinds of prey, and such specialised hunters cannot consistently gather enough food to develop large and sophisticated colonies. Yet paradoxically it is this very characteristic—small communities with simple structures—that helps them to diversify and survive in a variety of environments.

Over fifty million years ago some ants began to alter their hunting and gathering strategies. Rather than simply killing and eating sap-sucking bugs, they learned to herd and 'milk' them, just as we herd and milk cattle and sheep. These ant-shepherds tend their flocks with utmost care, driving off insect predators, and, if the flow of sap on which the herd depends begins to dry up, the bugs are carried to richer pastures. The ants even construct shelters that they herd their charges into in bad weather. At around the same time that this Lilliputian pastoral society was coming into being, other ants took a different path. They learned that instead of killing rival ants they could take them as slaves. Different ants discovered that they could save themselves the task of bringing up their young by laying their eggs in the nests of other ant species, much like cuckoos do among birds. But, most important to our story, there were ants that discovered agriculture, and the tale of their evolutionary path illuminates more clearly than any the path taken by our own species. The agricultural ants are known as attine, or leafcutter, ants, and are doubtless familiar from wildlife documentaries—their orderly columns of workers carrying leaf fragments like processional flags. Attine ants are found only in the Americas, and have been described by ant experts Bert Hölldobler and E. O. Wilson as 'Earth's ultimate superorganisms'.[77]

Attine ants are divided into castes—workers, soldiers and queens—which, as Eugene Marais found, equate in function to our organs. Much like our blood or skin cells, worker ants are short-lived, with 1 to 10 per cent of the entire worker population dying each day; in some species nearly half of the ants that forage outside the nest die each day. Queen ants, however, are longer lived, and can lay twenty eggs every minute of their decade-long lives. To continue the analogy, the soldiers are like weapons of offence or immune systems. And, as with our bodies, parts of an ant superorganism are made up of non-living matter. In our case it's our skeletons and the dead outer layer of our skin, but with ants it's the nest itself that provides structure and protection.

Some insect superorganisms are close in size to elephants and whales. Those of one species of leafcutter ant from South America can accomplish the excavation of forty tonnes of earth. Such giant colonies comprise up to eight million individual ants, and coordinating the action of so many individuals is one of the superorganism's greatest challenges. As we've seen, the principal tools used to achieve this are pheromones, potent chemicals that are so pervasive and sophisticated that it's appropriate to think of ants as 'speaking' to each other through them. Around forty different pheromone-producing glands have been discovered in ants and, although no single species has all forty, enough diversity of signalling is present to allow for the most complex interactions. The fire ant, for example, uses just a few glands to produce its eighteen pheromone signals, yet this number, along with two visual signals, is sufficient to allow its enormous and sophisticated colonies to function.

Pheromone trails are laid by ants as they travel, and along well-used routes they take on the characteristics of a highway. From an ant's perspective they are three-dimensional tunnels, perhaps a centimetre wide, that lead to food, a garbage dump or home. If you wipe your finger across the trail of ants raiding your sugar bowl you can demonstrate how important the pheromone trail is: as the ants reach the break in their trail they will become confused and turn back or wander around. The chemicals used to mark such trails are extraordinarily potent. Just one milligram of the trail pheromone used by some species of attine ants to guide workers to leaf-cutting sites is enough to lay an ant superhighway sixty times around the Earth.

While pheromones allow ant colonies to behave in 'intelligent' ways, theirs is an intelligence of a very particular kind. No ant carries around a blueprint of the social order in its head as we do, and there is no overseer or 'brain caste' that regulates the colony's activities. Instead, ants create strength from weakness by pooling their individually limited capacities into a collective decision-making system that bears an uncanny resemblance to our own democratic processes. This capacity is most evident when an ant colony has reason to move. Many ants live in cavities in trees or rocks, and the size, temperature, humidity and precise form and location of the chambers are all critically important. Individual ants appraise new cavities using a rule of thumb known as Buffon's needle theorem. Laying a unique pheromone trail across the cavity, each ant walks about the space for a period of time. The smaller the cavity is, the more often each one crosses its own trail.[78] Buffon's needle theorem yields only a rough measure of the cavity's size, and some ants may choose cavities that are too large or too small. The cavity deemed most suitable by the majority, however, is likely to be the best, and the way ants 'count votes' for and against a new cavity is the essence of elegance and simplicity. The cavity visited by the most ants has the strongest pheromone trail leading to it, and by following this trail the superorganism makes its collective decision. The band of sisters sets off with a unity of purpose, dragging their gargantuan queen and all their eggs and young to a new home that will give them the best chance of a comfortable and successful life.

Attine ant colonies are made up of a single queen and millions of workers that vary in size and shape (the largest being two hundred times heavier than the smallest). This allows for a sophisticated division of labour among the workers that rivals the kind of specalisation seen among human workers in complex industrial factories. The largest ant workers travel far from the nest to climb rainforest trees, and there they cut pieces from leaves which they drop to the ground. Other ants collect the cut fragments and carry them to a depot. From there, depot workers carry the cached fragments to the nest, where smaller ants undertake the process of turning this raw material into useful products. This work begins with middle-sized workers that cut the leaf fragments into smaller pieces. Smaller ants take these pieces and crush and mould them into pellets, which even smaller ants then plant with strands of fungus. Finally, the very smallest worker ants, known as minims, labour on these seeded pellets, weeding them and tending the growing fungus. These minute and dedicated gardeners do, however, get an occasional outing, for they are known to walk to where the leaves are being cut and hike a ride back to the nest on a leaf fragment—their purpose being to protect the carrier ants from parasitic flies.

The parallels between the ants and ourselves are striking for the light they shed on the nature of some everyday human experiences. In the superorganism, some ants get forced into low-status jobs and are prevented from upward mobility by other members of the colony. Garbage-dump workers, for example, are confined to their humble and dangerous task of removing rubbish from the nest by other ants who respond aggressively to the odours that linger on the dump workers' bodies.[79] Some of the most fascinating insights into ant superorganisms have come from studies of the amount of CO_2 given off by colonies. This is rather like measuring the respiration rate in humans in that it gives an indication of the amount of work the superorganism is doing. Researchers discovered that colonies experiencing internal conflict between individuals seeking to become reproductively dominant produce more CO_2 than do tranquil colonies where the social order is long established. But they also discovered that following the removal of a queen ant CO_2 emissions from a colony abruptly drop and remain low for three hours.[80] While it's not always helpful to anthropomorphise, it seems that ants may have their periods of mourning, just as we humans do when a great leader passes from us.

Despite these parallels, ants are clearly fundamentally different from humans. A whimsical illustration concerns the work of ant morticians, which recognise ant corpses solely on the basis of the presence of oleic acid, a product of decomposition. When researchers daub live ants with the acid, despite the fact that they are manifestly alive and kicking, they're promptly carried off to the ant cemetery by the undertakers. Indeed unless the daubed ants wash themselves very thoroughly they are repeatedly dragged to the mortuary, despite showing every other sign of life.

A single attine ant colony has about the same number of inhabitants as did England in the sixteenth century, but we should not think of either human or attine superorganisms as comprising solely humans or ants. Both consist of multiple species in intimate coexistence. And because the other species are 'owned' and controlled by the ants or humans, they are usually managed sustainably. Attine ant agriculture, however, is more sophisticated than many human agricultural practices, which are after all more recent. The fungus the ants grow, for example, has been cultivated by them for so long that it now exists nowhere else. The same is true of specialised bacteria that produce fungicides to destroy competing fungi and which are found only in special pocket-shaped crevices on the ants' bodies. In terms of agricultural sustainability, the ants provide an enviable model.

The insect superorganisms continue to evolve. Fire ants have recently formed a superorganism that has a similar geographical spread to that of a modern human nation. It came into existence just eighty years ago, yet it already covers most of the southern US and consists of billions of individuals. The founding fathers of this New World civilisation arrived in Mobile, Alabama, from somewhere in South America in the 1930s. Perhaps their *Mayflower* was a merchant ship carrying timber or other produce. Whatever the case, once they disembarked they set about colonising their new-found land with an energy that even the pilgrims would have found hard to match. Within fifty years their frontier had almost reached the limits of the land that was habitable for them: Virginia in the east and Oregon in the west.

In their native homeland fire ants form discrete colonies, with just one or a few queen ants at each centre. This is how most ants live, but something very strange happened to the fire ants in North America. They gave up founding colonies by the traditional method of sending off flights of virgin queens, and instead began producing many small queens, which spread the colony rather in the way an amoeba spreads, or the way suburbs extend a city, establishing extensions of the original body. Astonishingly, at the same time the ants ceased to defend colony boundaries against other fire ants. Ant populations thus coalesced into an intercompatible population spread across more than a million square kilometres in the US alone.[81] If individual attine ant colonies can be thought of as analogous to small nations, then the fire ants of North America resemble a federation, through which an individual ant could theoretically wander unchallenged from Virginia to Oregon. The cause of this remarkable transformation, the geneticists tell us, was a change in the frequency of a single gene, suggesting that the most far-reaching shifts in social organisation can by caused by seemingly small triggers.

Is it possible that human cultural evolution is driving us in a similar direction? The invention of the internet, mobile phones and cheap air travel are dramatically challenging the boundaries and capabilities of the old human superorganisms—the nations.

When viewed in the fullness of geological time, humanity's rise from hunter-gatherer societies to twenty-first-century civilisation appears instantaneous, so swift that a million years from now little or no evidence of it will remain in the geological record. Yet over that time we've gone from a population of around four million to nearly seven billion, and from a state of organisation where the largest functioning unit was the clan—similar in size and structure to a pride of lions—to one where our largest functioning unit includes almost every human being on Earth. Future beings pondering these astonishing events could look to the ants for some of their answers, for they will endure regardless of what happens to us, and it is in their complex and varied societies that keys to unlocking humanity's self-domestication may be found.

CHAPTER 10

Superorganismic Glue

During the time men live without a common
power to keep them all in awe, they are in
that condition which is called war; and such
a war as is of every man against every man.
THOMAS HOBBES 1651

An enormous field of research known as complex adaptive theory seeks to explain how simple elements self-organise into complex entities such as superorganisms. Much of the research is mathematical in nature.[82] Here I'd like to concentrate on just a few aspects of it. We have a sense that fortune smiled on the ants as they advanced towards superorganism status. It's generally understood that a colony's worker ants are each other's closest relatives and that this somehow makes for a society where individuals put the interests of their civilisation before their own.

Bill Hamilton wrote a precise mathematical formula that explained this—Hamilton's rule shows the circumstance in which individuals are likely to make significant sacrifices for others. It can be expressed as $C < R \times B$ (C being the potential reproductive loss of the one making the sacrifice, R being the closeness of the genetic relationship between the two, and B being the benefit the recipient derives).[83] The multiplying effect of relatedness is clearly critically important, individuals that share many genes in common being more likely to sacrifice for each other. Hamilton's rule has wide implications, telling us why, for example, worker ants don't reproduce but relegate that duty to their queen, and why a parent might, at the risk of losing their life, jump into the sea to save a drowning child.

If Hamilton's rule were our only hope for a peaceful, cooperative society, we might as well go and hang ourselves, for the social glue the rule provides is so dependent upon relatedness that it's barely strong enough to hold together an extended human family, much less a multicultural society. In the real world—even among the ants—things are not as simple as this splendidly crisp formula suggests and, to see why, we need a closer look at those six-legged marvels.

All worker ants are female, and they tend to share a very high degree of genetic relatedness. This is particularly true among species in the early stages of superorganism development. Their queens are strictly monogamous—mating with a single drone on the nuptial flight (and storing the sperm for a lifetime of use)— which ensures that all workers have the same parents. Things are different, however, among the most highly evolved ant superorganisms. Queen attine ants, for example, mate with several males during their nuptial flight; thus, workers have different fathers. This greater degree of genetic variability confers considerable advantages on the colony, including resistance to disease and a wider variety of behaviours and body types among workers, but it inevitably weakens the bonds elucidated by Hamilton's rule.

So, Hamilton's rule explains the early stages of superorganism development, but not its more complex flowerings. Why don't advanced superorganisms collapse under the pressure of their own genetic diversity? Clearly, another kind of superorganismic glue must exist.

It's a remarkable fact that, despite our huge population, humans are one of the most genetically uniform of mammal species, there being more genetic diversity in a random sample of about fifty chimpanzees from west Africa than in all seven billion of us.[84] All such genetically limited species hide a tragedy in their family trees—a brush with extinction, known as a genetic bottleneck, in which the population was reduced to a tiny size over many generations. Our own species' near-death experience, which occurred around seventy thousand years ago, may have been caused by the eruption of the Toba volcano in what is now Indonesia. Based upon studies of the 1991 eruption of Mount Pinatubo, it's estimated that Toba altered our atmosphere so dramatically that the average surface temperature of Earth dropped by 2° to 3° Celsius, killing all but between one thousand and ten thousand breeding pairs of humans. While this did not reduce our genetic diversity to the point where Hamilton's rule might see us cooperate with all, its legacy on our culture is clear, for we share many characteristics and much understanding—an essential humanness, if you like.

The commonality of human understanding was brought home to me when I conducted biological surveys in remote parts of New Guinea. Occasionally I encountered people who had not met an outsider before. My ancestors and theirs had parted ways at least fifty thousand years ago, when each band trooped out of Africa to its different destinations. Yet when we met, after fifty millennia of separation, I understood instantly the meaning of the shy smile on the face of the young boy looking at me, and he understood my motion for him to step closer to better observe what I was doing. There was much natural magic to those unforgettable meetings. I often sat down at night to food prepared by a lonely widow, and in eating we shared the basic human need to care for each other. When I pantomimed to hunters that I wished to learn about their animals, I was immediately understood. The men vied with each other to pass on details, in imitated sounds and mime, of the creatures of their world. And slowly, as we learned a few shared words, those people came alive to me as individuals.

Despite this commonality, we've been very good at living as if our family, our clan or our nation is the only truly civilised and 'proper' group of people on Earth, and believing this has enabled us to kill and rob and maim each other without seeing that we are thus damaging ourselves. Nothing is as challenging to such a belief as meeting the 'other' on an equal footing. There's as much diversity of thought, mannerism and emotion in a small New Guinean village as there is in the entire world, and in this commonality lies the foundations of our universal human civilisation, as well as its hopes for a future.

The glue that holds complex superorganisms together is not genetic, but social—yet it is enhanced by a sufficiently narrow genetic base to enable universal understanding. It was the eighteenth-century economist Adam Smith who first elucidated its nature and magic—a magic that allows the prodigious productivity of our industrial societies. He called it 'the division of labour' and in seeking to explain it he asked us to consider the pin factory that served as his exemplar. A pin, he says, may be a simple object, but its manufacture requires up to eighteen separate operations—cutting the wire, pointing the pin, fixing the head to it and so forth—and in a large pin factory each operation is performed by a specialist who does nothing else. Smith doubted that a person working alone and performing all eighteen operations could make twenty pins in a day. Yet he found that ten men specialising in just one or two operations could make forty-eight thousand pins per day.

This near-miraculous increase in output Smith saw as the ultimate source of the prosperity of advanced nations.[85] But the roots of the division of labour are far more ancient than pins. They go back to our very beginnings as a species.

Ever since men and women began to do different tasks, the division of labour has benefitted our species. It allowed camps to be provisioned and protected, and children nurtured. It took a mighty leap twelve thousand or more years ago, when some lonely person, perhaps, took the first wolf into his camp. Can you imagine what it must have been like to sleep well for the first time, secure in the knowledge that the wolf's sharp nose and ears would hear an approaching predator first; or to snuggle into its warm, furry body as the frost of night descended? And the wolf? It too got warmth, some scraps and a superior hunting pack. The division of labour thus created was enormously productive. A dog can track a creature that no human could hope to find, but it often takes a human to catch and subdue it. Such experiences led to the civilising of ourselves, and, as we added one species after another to the miniature ecosystems that constitute our communities, we discovered that each creature brought marvellous benefits, as well as a host of hidden costs.

As our interdependence has grown, we've shaped our animal companions to suit our needs. It's convenient that male mammals bear their testicles in a pouch on the outside of their bodies, for our ancestors soon learned how to rob young males of their reproductive treasure and so fatten them more quickly. They learned also how to select only the finest, most tractable males to father the next generation, and so they shaped dogs, cats, cattle and sheep—even creatures whose anatomy was not so conveniently arranged—into what they are today. We in effect became the most powerful of evolutionary forces, bending natural selection to our own ends, and so creating creatures the likes of which had never existed before and which could survive only in a miniature ecosystem of our own making.

While we may think occasionally of how we have shaped animals such as dogs and sheep, we dwell less frequently on how much we ourselves have changed over those ten thousand years. But change we have, and in the most remarkable ways. Archaeologists studying the earliest stages of domestication see the effects of this process on our ancestors, including a steady increase in human numbers and an ever more settled lifestyle. But they also find evidence for a stunting of the human frame caused by an inadequate diet, and a terrible mortality and crippling of bodies brought about by new diseases. It must have taken our ancestors some time to understand that grain and a little meat were not sufficient for a healthy body—that those greens, nuts and other delicacies gathered on the seasonal round of the nomad were essential to good health. And it must also have taken time to learn that huts must be kept clean and livestock housed separately if disease was to be avoided. We know from observing modern hunter-gatherers who are trying to adopt a modern lifestyle that such lessons are difficult to learn. So it was that the likes of smallpox, chickenpox, measles and anthrax decimated our ancestors.[86]

But these were not the only costs affecting those who sought to sup from the cornucopia that is the division of labour, for it also diminished our finest faculties, eroding our individual independence and mental acuity. And therein lies a paradox—one which is shared with the ants—that while agricultural societies are powerful, they are composed almost entirely of incompetent individuals.

To gain the meaning of this in full measure, just compare a day in your life with that of a hunter-gatherer such as an Australian Aborigine. On rising each morning Aborigines must find and catch their own food, make or repair their tools and shelter, and defend and educate their families. They are thus their own provider, manufacturer and protector. Put in an Aborigine's place, we'd be as lost as white rabbits in the wilderness; our tenure in the world most likely counted in days rather than months.

The reverse, however, is not true. History shows that hunter-gatherers can learn to do any of the jobs our society offers. I've flown in a helicopter piloted by a New Guinean who was born into a traditional society all but innocent of metal. And history is replete with examples of academically gifted Native Americans and Aborigines—like John Bungaree, who topped the class in mathematics, geography and writing in early-nineteenth-century Sydney.[87] There are even a few examples of hunter-gatherers giving farming a try. But regardless of their accomplishments, almost all of them went back to their own culture. The truth is that hunter-gatherers find the loss of liberty we routinely endure to be insufferable. The rules we obey simply to sit on a train, or walk along a street, may be second nature to those born into such straitjackets, but to others they are monstrous. And for one used to being his own provider, warrior and law enforcer, our daily round is interminably boring.

Yet what has happened, time and again, when we feeble cogs in the mechanisms of complex societies meet with superbly competent hunter-gatherers? It is the poxed, incompetent weaklings who triumph, leaving the splendidly strong and well-nourished bones of the hunter-gatherer in the dust.

This tendency towards civilised imbecility has left its physical mark on us. It's a fact that every member of the mini-ecosystems we have created has lost much brain matter. For goats and pigs it's around a third when compared to their wild ancestors. For horses, dogs and cats it may be a little less.[88] But, most surprising of all, humans have also lost brain mass. One study estimates that men have lost around 10 per cent, and women 14 per cent of their brain mass when compared to ice-age ancestors.[89] It's easy to see why. The dog's sharp nose protects the sheep from danger, while the herder's knowledge of pasture means that the sheep don't even have to think about where they'll forage for the day. And of course the wether's bones and other scraps relieve the dog from having to hunt, while its meat and skin feeds and shelters the man. Overall, life for all members of our domesticated mixed feeding flock is made so much more accommodating that its members can invest less of their energy in brains and more in reproduction and fighting disease.

The ultimate point to which the division of labour had driven humanity by the close of the eighteenth century was remarked upon by Adam Smith, who said of the repetitive and narrow work in the pin factory that:

> The man whose whole life is spent in performing a few simple operations...generally becomes as stupid and ignorant as it is possible for a human creature to become. The torpor of his mind renders him not only incapable of relishing or bearing a part in any rational conversation, but of conceiving any generous, noble, or tender sentiment...Of the great and extensive interests of his own country he is altogether incapable of judging; and unless very particular pains have been taken to render him otherwise, he is equally incapable of defending his country in war.[90]

While Smith may have overstated the case, not least in his dismal view of the worker's potential to improve himself, his general point should serve as a warning to us. As Samuel Butler wrote, 'Every man's work...is always a portrait of himself.'[91] If you doubt how far our civilisation has turned us into helpless, self-domesticated livestock, just look at the world around you. Does it seem to have lost its commonsense? It frequently seems so to me. And how often does a visionary leader arise among us? So few are truly wise that it seems a whole generation can pass without such a presence. But of course it need not be so. We can challenge ourselves—shed our complacency and love of ease—and so reinvigorate our shrivelled virtues. That's what a well-rounded education is supposed to do, and even those with the most repetitive jobs can in their leisure hours expand their minds. But while we sit in our air-conditioned homes and eat, drink and make merry like cattle in a feedlot without the slightest thought about the consequences of our consumption of water, food and energy, we only hasten the destruction—in the long term—of our kind.

The division of labour delivers one great good in that it provides the glue for our civilisation. It is the glue of trade, a process which makes both parties wealthier than they were before, and the force it must supplant is that of the warrior with an imperative simply to take. An indication of the strength of this glue can be found by asking what resource draws people to the metropolis. It is, of course, each other—for there is no other worthwhile resource in a city—and so powerful is its pull that for hundreds of years cities have been population sinks, places where mortality, driven by appalling sanitation, crowding and pollution, consumed the endless stream of humanity drawn to them from the healthier countryside. But still they came, and despite the crowding, high prices and pollution they continue to come, because to sup from the table of labours divided is so valuable that it's even worth risking death for. And once our voluntary surrender to the competencies of others has gone so far, we find that we simply cannot exist outside a civilisation.

The division of labour thrives where there is peace and security, and is imperilled where crime and civil strife reign. Among hunter-gatherer ants internal conflict can be ceaseless, but in the gigantic colonies of attine ants civil strife is unknown. No doubt the attine ants have gone so very far down the path to mutual interdependence, that, to quote Macbeth, 'returning were as tedious as go o'er'. The attine colonies, however, remain willing to prosecute wars against their neighbours as well as to defend themselves against external attack. As we examine war and peace in human societies, it's important to discriminate between two kinds of conflict: internal disputes such as civil unrest, civil wars and crime-related violence; and conflicts fought with outside entities.

In advanced human societies internal strife is suppressed by government, which reserves to itself the right to violence. But while humanity remains divided into competing power blocs, conflict between those blocs will continue to occur. And the more power blocs there are, the more chronic the state of war. It's thus impossible to chart humanity's pursuit of peace and war without reference to the size and complexity of the political entities involved. Prior to the agricultural revolution, everybody lived in family-sized clans, each led by an adult male or a group of men, and it was they who administered justice. But by the eighteenth century the clans had cohered into hundreds of nations and smaller states, the largest only tens of millions strong, and by the mid-twentieth century nearly all of humanity had formed into two opposing blocs, each of which had the power to destroy the planet.

What do we know of warfare and conflict in clan-based societies? At an anecdotal level, a stroll round any museum will give you some idea. I recently visited the Danish National Museum, where dozens of skeletons of Neolithic people are on display. Almost all, it seemed, whether male or female, adult or child, had suffered a blow to the head, and some had been scalped. The tools used for the banging and scalping were everywhere. Indeed, weapons were invariably the most ornate and lovingly crafted pieces of kit produced by ancient cultures. The American evolutionary biologist Samuel Bowles has systematised this impressionistic answer by amassing evidence for physical violence from ancient graves (of both farmers and hunter-gatherers) as well as from surviving hunter-gather groups. Of the skeletons excavated from graves, up to 46 per cent showed signs of a violent death, and life among other hunter-gatherer societies, it seems, was hardly less severe. Among the Ache people of eastern Paraguay, for example, 30 per cent of deaths occurred as a result of violence in war, while among the Hiwi of Venezuela and Colombia, 17 per cent died the same way. Among the Tiwi, an Aboriginal group from northern Australia, the total was 4 per cent.[92]

Hunter-gatherer societies lack police and jails, and in some there are few goods that can be appropriated by way of a fine. As a result there are no means to punish transgressors except for banishment (which weakens the clan as a whole) or corporal punishment. Just how this results in increased violence can be glimpsed from observations made when literate societies encounter hunter-gatherers for the first time. Few are more detailed and telling than those concerning the Aboriginal population of the Sydney area at first contact in the late eighteenth century. Watkin Tench, a lieutenant in the Marines, wrote this about Gooreedeeana:

> She belonged to the tribe of Cameragal and rarely came among us. One day, however, she entered my house to complain of hunger. She excelled in beauty all their females I ever saw. Her age about eighteen, the firmness, the symmetry and the luxurancy of her bosom might have tempted painting to copy its charms...I was seized with a strong propensity to learn whether the attractions of Gooreedeeana

were sufficiently powerful to secure her from the brutal violence with which the women are treated, and as I found my question either ill-understood or reluctantly answered, I proceeded to examine her head, the part on which the husband's vengeance generally alights. With grief I found it covered by contusions and mangled by scars. The poor creature, grown by this time more confident from perceiving that I pitied her, pointed out a wound just above her left knee which she told me was received from a spear, thrown at her by a man who had lately dragged her by force from her home to gratify his lust.[93]

Males also bore the marks of physical violence. One of Tench's firmest Aboriginal friends was Bennelong. When Tench first saw him, in December 1789, he was 'about twenty-six years old, of good stature and stoutly made, with a bold intrepid countenance which bespoke defiance and revenge'.[94] Just a few months later Tench noted that:

He had received two wounds in addition to his former numerous ones since he had left us; one of them from a spear, which had passed through the fleshy part of his arm; and the other displayed itself in a large scar above his left eye. They were both healed, and probably were acquired in the conflict wherein he had asserted his pretentions to the two ladies [his new wives].[95]

Bennelong received severe wounds again in 1796, including a blow that divided his upper lip and knocked out two teeth, and he endured numerous other spearings and blows before his death in 1813.

I have chosen these examples, which are possibly not entirely typical, because they provide relatively detailed individual records.

Today few people bear even a fraction of the physical damage suffered by Bennelong or Gooreedeeana. Indeed few die from violence at all, even in the gun-toting US, which has the highest homicide rate in the developed world (around 5.4 per hundred thousand per year).[96]

While internally peaceful, nations can engage in conflicts with other nations that are far bloodier that anything prosecuted by hunter-gatherers. Australia was a minor player in World War I, yet it sent four hundred thousand of its three million males to fight, and from the Somme to Gallipoli some sixty thousand Australians died. It's a horrendous toll, yet it represents just 1 per cent of the total population of Australia as it was at the time. The death toll in Europe was also devastating, but as the Bishop of London said in 1917, an average of nine British soldiers died every hour during 1915, while twelve British babies died every hour through that same year. In other words, infant mortality remained a greater cause of British deaths than violence even during the Great War.[97] You might think that World War II was a more devastating conflict, and indeed the death rate in the German military was among the highest ever recorded for any developed nation. If calculated on the basis of Germany's population within its 1937 borders, the percentage of people killed directly in military conflict was 6.2.[98]

While our world has changed since 1945, mutually assured destruction remains a threat. What will happen as the human superorganism becomes global? While it's difficult to determine cause and effect, the trend in the era of globalisation has been towards fewer deaths in war. The Peace Research Institute, Oslo (PRIO) estimates that between 1946 and 2002, the annual battle toll around the world has declined by more than 90 per cent. This finding has implications for defence policy, and US researchers have argued that the toll in recent decades was in fact nearly three times higher than PRIO's estimate. PRIO has defended its estimates, however, and a study published in the *Journal of Conflict Resolution* has determined that the US team was making different measurements from PRIO.[99] The PRIO estimates prompt two questions: When will the battle toll reach zero? Is it possible that humanity might see an end to war between nations?

We cannot answer those questions without considering the political structures our species has evolved. In the Stone Age the politics of the clan were simple and also recognisable in numerous other mammal species: a family structure at whose centre sat a dominant male, one or several mature females and their offspring. The first villages must have consisted of clusters of such units. But what happened after that? Speculation on the progress of human societies is at least as old as Plato's *The Republic*, which begins by outlining the characteristics of an ideal society. In it, Plato says, men and women would do similar work, and all reproductive effort would be held in common, the women and children being tied to no individual. The reason for this, Plato says, is that 'the best ordered state is one in which as many people as possible use the words "mine" and "not mine" in the same sense of the same things... What is more, such a state most nearly resembles an individual'.[100] What Plato is describing here, some two and a half thousand years before modern science, are the principles upon which an ant colony works. Plato recognised that humanity would not take readily to such a system, but instead of giving it up as impractical he suggests a program of eugenics to create a race more amenable to it. A band of old men, he thought, could manipulate the opportunities for copulation so skilfully that the population as a whole would be unaware of it. They could do so, for example, by holding festivals at which sexual licence would be given to certain couples if it was felt the offspring would further the interests of society.

As things turned out, our species hit upon another schema to order its societies—one which is entirely inimical to Plato's solution. Called the democratic process, it puts the individual and his or her will front and centre. Plato had much to say about it—all under the heading of 'imperfect societies'—and democracy must be classified in this way when compared to the societies of ants, which perform like pieces of well-oiled machinery. But democracy is uniquely suited to the ordering of societies of willful and self-centered apes; as Winston Churchill said of it:

> Many forms of government have been tried, and will be tried in this world of sin and woe. No one pretends that democracy is perfect or all wise. Indeed, it has been said that democracy is the worst form of government except all those others that have been tried from time to time.[101]

Plato argues that democracy arose from a situation where all political power rested with land-owners, a system he calls 'timocracy'. The transition, he thought, came about as a result of lack of constraint in the pursuit of getting as rich as possible, which he saw as softening the rich and strengthening the poor whom the rich hired to do many tasks, such as fighting, on their behalf. But democracy in its turn, Plato believed, must give way to tyranny—for the tyrant rises as a popular champion, and democracies lack the means of restraining such individuals.[102] The kinds of democracies that existed in the ancient world were very different from those of our era. They permitted the holding of slaves, and those with the right to vote were a select few indeed, principally powerful men.

Only in the twentieth century has democracy lived up to its name, encompassing all adult members of a society. And as it has done so it has provided a more powerful superorganismic glue than any before—self-interest. That's because the rights of the individual are inextricably bound up with the democratic process, and those rights include protections for those wishing to keep the benefits of their labour. Looking at the spread of democracy in the modern world, it's tempting to think that it has now found the strength to resist tyranny.

If the gulf separating Plato and modern democracies is large, it encompasses just a tiny part of the history of human superorganisms. It's now time to turn to the development of the five human superorganisms, and what they can teach us.

CHAPTER 11

Ascent of the Ultimate Superorganism

When tillage begins, other arts follow. The farmers, therefore, are the founders of human civilisation.
DANIEL WEBSTER 1840

What, in essence, is a human superorganism? Insect superorganisms result from a direct expression of a genetic inheritance. While we humans may be built by our genes, our civilisations are built from ideas. Thus, at the heart of a human superorganism is a mneme—a complex of ideas—about how to use other species for our benefit, which can be passed on to others culturally. So we gather to ourselves the genes required, whether they be in dogs, cattle, grasses or other people, in order to form a miniature, artificial ecosystem, at the heart of which a city often develops.

The growth of human superorganisms is far more mysterious than the evolution of individual species, or even insect societies. Civilisations expand, divide, borrow and collapse in ways that often seem inexplicable from a purely evolutionary perspective, in part because the temperaments and capacities of powerful individuals can have profound effects on the course of history. There are nevertheless rules that guide the development of civilisations, and by examining the origins of agriculture, and what grew from them, we can learn something of those rules. We can also learn much from what happens when one superorganism comes into contact with another.

The stimulus for humans to settle down and grow crops was almost certainly a change in climate, a shift from the unpredictable and hostile climate of the ice age to a period of remarkable climatic stability. If this was indeed the magic ingredient that seeded civilisation, then humanity must have an inherent propensity—which was frustrated for millennia by an adverse climate—to superorganise. We know little of the details of the ice-age climate regime, but it seems that its vicissitudes must have been so great, and its swings so perilous, that it was dangerous for our ancestors to settle in one place and to rely on a crop for a year's supply of food. They had to keep moving on, carrying their tools with them, as they exploited one resource after another.

Agriculture arose independently in five regions: the Fertile Crescent (an arc of land extending from Turkey to Iran), East Asia, South America, North America and New Guinea.[103] Each region can be thought of as giving rise to a separate human superorganism, and each presumably began with a few experiments with sowing and tending crops, and domesticating animals. Over time, as agriculture spread, new animals and plants were included, and local variants of already domesticated types added. So it is, for example, that both European and Indian varieties of cattle were brought into our herds.

The most surprising early agricultural development is surely New Guinea. In *Guns, Germs, and Steel*, which appeared in 1997, Jared Diamond said he could not be certain whether New Guinea represents a truly independent origin of agriculture, because, when he wrote the book, the kinds of crops grown were unknown. But more recently they've been identified as taro and banana.[104] Recent studies also demonstrate that the New Guinean superorganism is one of the most ancient, dating back more than ten thousand years. It is also arguably the most productive, providing a resource base for the highest rural population densities on Earth—1614 people per square kilometre.[105]

Taro is a highly nutritious root crop with 'elephant-ear' leaves. It grows in swamps, and to plant it you need only slice off the green top and thrust it into water-logged soil. The quality of taro tubers varies greatly—some are too astringent to eat, while others are delicious. I can imagine a woman (it's almost invariably women who gather plants and tend gardens) selecting the leafy top of a particularly delectable taro tuber enjoyed by her family the night before, and taking a few seconds to place it back in the soil. When she and her family returned to the spot months later, there would be more taro to eat. Over time, gardens were formed and people began to drain areas that were too wet for the plants to thrive, as well as to protect crops from animals such as rats and possums, which are also fond of taro. Doubtless they used stone axes (which have been made in New Guinea for forty thousand years) to fell trees that threatened to shade the garden. Today at least six hundred varieties of taro are grown in New Guinea. Among the most prized is a red one that tastes sublimely buttery. So valued is it that in the Telefomin area a poor person is referred to as a man without red taro.

The cooking banana, also first domesticated in New Guinea, is not widely appreciated outside the tropics, but in many parts of Africa, South America and Asia it's a staple, taking the place of potatoes. For all their value, however, neither taro nor cooking bananas can compete in global importance with sugarcane. The two most widely planted species are New Guinean in origin, and in first domesticating them some anonymous soul gave humanity its most economically valuable crop. And, in a world on the brink of a biofuel revolution, its importance is only set to grow, for sugarcane is our best source of ethanol.

Despite its early and promising start, the New Guinean superorganism developed no cities, or indeed stone buildings. Its people continued to live in villages of a few dozen huts, its leaders differing little from the head of a band of hunter-gatherers, right up to the time of European contact in 1933. And they remained a Stone-Age people, despite the fact that entire mountains of copper as well as rich gold deposits existed on the island. Jared Diamond speculates that the lack of domesticable animals was an important reason, but I'm not so sure. Archaeological evidence shows that by ten thousand years ago New Guineans were introducing wallabies and possums to islands lacking game animals, and to do that they must have at least commenced the domestication process. Even today tame possums and tree-kangaroos abound in New Guinean villages, and, while there may have been some fatal impediment to their further development as domestic stock, it's a remarkable fact that the Javanese bred domesticated wallabies—obtained from the New Guinean region—for food, which is how the Europeans first encountered them.

The human superorganisms of the Americas got off to a far later start. Experiments with agriculture commenced between four thousand and five and a half thousand years ago.[106] Potatoes, sweet potatoes, maniocs, tomatoes and other important crops originated in the valleys of the Andes, coastal Peru and perhaps the adjacent Amazon. Llamas and guinea pigs were domesticated in those places too. Corn, beans and squashes, among other crops, were first domesticated in what is today Mexico and Central America, while a second focus of agriculture involving different crops such as sunflowers occurred in what is now the eastern US. This, however, developed only around two and a half thousand years ago, so it's possible that the idea of agriculture spread there from Mexico.

Both the North American and South American superorganisms gave rise to sophisticated, hierarchical societies that built cities, discovered metallurgy and made other breakthroughs—all in synchronicity. This is truly remarkable, as there is no evidence whatsoever that these farmers and builders had any contact with each other. Cities were established by 2900 years ago at Norte Chico on the Peruvian coast and by 2700 years ago at Tikul in central Mexico. By 2800 years ago metallurgy was being practised in what is now western Mexico and in coastal Peru. By five hundred years ago large empires had formed with impressive cities at their hearts. This parallel development is truly intriguing, and may point to something not yet understood about the tempo and mode of super-organism development.

For all their glory, the New World civilisations seem to have shared a characteristic that did not bode well. The buds that became their great cultural flowerings were prone to wither, and many of these cities were relatively short-lived. While a full expla-nation of this remains elusive, the particular climatic conditions of North America, with its extreme fluctuations, when combined with a limited resource base, were probably a key factor.[107]

The other two human superorganisms became established in Eurasia. Eurasia is by far the largest landmass, a whopping fifty-four million square kilometres, and it has a land connection with Africa, which comprises a further thirty million square kilo-metres. You can think of this landmass as consisting of three fertile peninsulas—Europe and the Middle East, India, and East Asia—

which are imperfectly separated by hostile plains, mountains and deserts. The two superorganisms that established themselves here—that of the Fertile Crescent and that of East Asia—were located at opposite ends of this vast land.

Rice and millet had been domesticated by 9500 years ago in the valley of the Yangtze and in northeast Asia. Permanent villages had become established by around 7000 years ago, and by 4100 years ago cities were developing and metallurgy was being practised. The centre of the East Asian superorganism has always been China. Early in its history, such as the 'spring and autumn period' between about 2491 and 2732 years ago, the region that is now China hosted hundreds of independent states. But since the establishment of the Qin Dynasty (221BCE), there has been a tendency towards the dominance of a single political unit. By seven hundred years ago China had become the largest and most prosperous state on Earth, and within a century it had developed a naval capacity capable of reaching Africa. Shortly thereafter, however, at the whim of an emperor, China turned inwards, abandoning its navy and thus its network of globe-straddling contacts.

China has been the point of origin for some of humanity's most important inventions, including gunpowder, printing, the compass and blast furnaces for iron smelting. These developments revolution-ised warfare and commerce, and by a thousand years ago they had given China paper money, cannons, land and sea mines, and military rockets. When, centuries later, these innovations reached Europe, their impact was dramatic. Sir Francis Bacon wrote of them:

> Printing, gunpowder and the compass: These three have changed the whole face and state of things throughout the world; the first in literature, the second in warfare, the third in navigation; whence have followed innumerable changes,

in so much that no empire, no sect, no star seems to have
exerted greater power and influence in human affairs than
these mechanical discoveries.[108]

The Fertile Crescent is home to humanity's oldest superorganism, with evidence of the domestication of dogs dating back
there twelve thousand years, and the cultivation of rye around
eleven thousand years. Towns (the oldest being Jericho) were established by around 11,500 years ago, and metallurgy by 5600 years
ago.[109]

Within eight thousand years of its origination, the superorganism of the Fertile Crescent had ramified and spread into the
furthermost reaches of Europe, deep into Africa and India, and east
into Eurasia as far as the silk-road oases of the Taklamakan Desert.
As it spread it blossomed into the civilisations of Mesopotamia, Egypt
and the Harappan culture of what is now India, as well as that of
classical Greece, to name just a few. These cultural flourishings
often overlapped in time, giving rise to exchanges, synergies and
conquests. But it was not until around two thousand years ago that
a culture arose that was, in some respects, recognisably modern,
and which would eventually occupy much of the Fertile Crescent's
heartland. That culture was the Roman empire, and it's worth
examining briefly for it represents a template for the modern world.

The engineering marvels of Rome capture our imagination.
The Baths of Caracalla—with their immense domes and glass
facades, their hot and cold water, their libraries, restaurants and
other entertainments—have no equivalent in the modern world.
Nor, until the twentieth century, had the Colosseum, which could
seat seventy thousand, or the Circus Maximus with its capacity
for a quarter of a million people. From the aqueducts of Rome to
its road system, to marvels of elegance such as the Pantheon, the

achievements of this civilisation continue to amaze. But even more astonishing is the extraordinarily contemporary nature of many aspects of Roman life.

At its height Rome was home to one and a half million people, yet it had fewer than two thousand free-standing houses. Most Romans lived in the city's forty-six thousand *insulae*, or apartment blocks. This made Rome a high-rise city whose density was unparalleled in the ancient world, and whose closest resemblance is to cities such as late-nineteenth- and early-twentieth-century New York. Emperor Augustus had decreed that *insulae* should rise no higher than twenty-one metres (about seven storeys) but breaches of the regulation were common. Nobody knows how tall the famous Insula Felicles was, but judging from contemporary accounts it was the Empire State Building of its day. In the absence of elevators or ducted water to the higher floors, the most desirable abodes in the *insulae* were on the first and ground floors.[110]

Mortality rates in the city were high and, as in the cities of early modern Europe, constant immigration was required to maintain the population. Africans, Egyptians, northern and southern Europeans and people from the Middle East could be seen every day. Just how dominant they became is revealed by one survey of names of Rome's inhabitants, in which 60 per cent or more of the names were of Greek, rather than Latin, origin. The Greek language was widespread—these people could have originated anywhere from what is today Greece, Syria, Turkey, Iraq or adjacent regions. The fact that at least six (and perhaps as many as eight) out of every ten inhabitants of Rome came from outside the Italian peninsula reveals a degree of multiculturalism not matched until modern times.

Another striking aspect of the Roman empire was the availability of fast food. There was a *popina* (fast-food stall) on almost every corner in Pompeii and Herculaneum, and it seems also in the city of Rome. Also modern was the equality of the sexes. By two thousand years ago, Roman women—particularly those from the leading families—were able to inherit goods, property and money, and to divorce their husbands with relative ease. Increasingly, couples were co-habiting for love, many not bothering about formal marriage at all. There was even a severe drop in the birthrate as women took control of their own fertility.[111]

There were, however, aspects of Roman life that look alien to us. It was a slave-owning society, and the savagery of the games held in the Colosseum are legendary. Yet I would argue that it's our technology that sets us apart from the Romans in these regards. We loathe slavery on moral grounds, but the functions once performed by slaves are oftentimes today performed by machines, and we have hardly weaned ourselves from our fascination with gore. Instead of seeing the real thing, however, we simulate it and broadcast it on television and in the cinema. There may, incidentally, be an evolutionary reason for this fascination. Birds will flock about if one of their number is caught by a hawk. Perhaps we're genetically predisposed to observe such things from a safe distance, because we may learn how to avoid a similar fate.

Rome has a particular fascination because it suggests that, once they develop to a particular level of complexity and sophistication, human civilisations tend to converge on each other in key ways. They extend many privileges to their citizens (even to slaves) regardless of sex, are characteristically multicultural and high in population density, and develop a kind of lowest common denominator cultural glue based on fast food and populist entertainment. They also have a remarkable resilience to changes at the top. Rulers can come and go, but the superorganism will continue. Their enemy is a diminution of resources; their Achilles' heel is interruption to the flow of energy—whether it be food or oil—or water.

Rome suffered a crisis when Christianised barbarians, led by Alaric the Goth, laid siege to the newly Christian city (pagan practices having been banned just twenty years earlier), stopping its food supply. No superorganism can long withstand such a diminution of resources, and the weakened city was sacked. Other sackings would follow, and in 439CE the food supply from North Africa was permanently cut. By 530CE, following the severing of the city's aquaducts, Rome, which at its height had a million and a half inhabitants, was reduced to a population of just fifteen thousand.[112]

Despite the ultimate collapse of Rome, some aspects of Roman civilisation never vanished. Its language and laws survived in various forms, for example. But the great infrastructure—the roads, the water-supply systems and the integrated reach for resources they facilitated—was not maintained. No comparable entity would arise again—neither the Holy Roman empire nor the conquests of Napoleon resemble in scale and longevity Roman rule. Indeed, it was not until the establishment of the European Union (EU) in 1993 that a sustained, similar-sized political entity would declare itself in Europe. At its height Rome, of course, was much larger than the EU, covering vast areas of Africa and the Middle East. It's unclear why it took so long for a such a large political entity to re-establish. But the experience should remind us that, should we be foolish enough to let our global civilisation break down, we should not expect a swift recovery.

One increasingly influential factor in the development of the civilisations of the Fertile Crescent was their interconnectivity with East Asia. Human superorganisms can take in goods and ideas from each other, and often the transfer of such ideas is enormously beneficial. Even before Roman times, goods and ideas were being traded along the silk road, and across the seas separating civilisations: cloves from the Moluccas (just west of New Guinea), for example, were making their way to Syria by around four thousand years ago. Despite such early trade, until the accounts of Marco Polo and other travellers such as Ibn Battuta, there was no real knowledge, in either superorganism, of the existence of the other. But as time went by ideas would travel ever more quickly. Galileo's book *The Starry Messenger*, which describes features of the heavens as seen through his telescope, was published in Europe in 1610. Just five years later it was available in China, in Chinese translation.[113] This partial connectivity contrasts with the isolation of the civilisations of the Americas and New Guinea. And that isolation was to have a terrible cost when the civilisations of the Fertile Crescent reached American shores.

By the time of Columbus the American superorganisms were in full flower. The Aztec city of Tenochtitlan was home to around two hundred thousand people, with perhaps a million living in the surrounding valley. It was a grand metropolis built in the middle of a lake, with three causeways linking it to the main-land. The Spaniard Bernal Díaz del Castillo was part of Cortez's army of conquest. The only European to leave a first-hand account of its discovery, he recalled that, when the Spanish first saw the city in November 1519:

> We did not know what to say, or if this was real that we saw before our eyes. On the land side were great cities, and on the lake many more. The lake was crowded with canoes...Some of our soldiers had been in...Constantinople, in Rome, all over Italy, but none had seen anything like the magnificence of Mexico.[114]

As Jared Diamond has memorably recounted, it was guns, germs and steel that saw to the destruction of this city, and indeed all the New World civilisations. So grand was the scale of devastation unleashed that it was never, in all of European colonial history, repeated. Within six months of the Spanish laying siege to the city early in 1521, a third of the inhabitants were dead, mostly from disease. Overall, it's been estimated that diseases such as smallpox killed between 90 and 95 per cent of the population of the New World.[115]

As one-sided as the devastation was in this clash of civilisations, the exchange that ensued was to enrich the entire world. Potatoes became a staple for much of Europe, supporting an expanded population, while in East Asia sweet potatoes became an important crop on lands too poor to support rice. And in the Americas the re-introduction of the horse (after its extinction there thirteen thousand years earlier) and cattle would revolutionise New World cultures. Most important from a global perspective, however, was the fact that, within just a few decades, the civilisations of the Fertile Crescent had annexed forty million square kilometres of land. That's 28 per cent of all the habitable surface of the Earth, much of it fertile and resource-rich. The great colonial expansion of Europe had begun.

The conquest of the Americas marks a decisive moment in the fate of the five human superorganisms. Two—those of North and South America—were vanquished and incorporated into that of the Fertile Crescent, and the wealth thus aquired would prove fundamental in fuelling further expansion. But also important was the spread of the idea that Europeans could colonise distant lands and expect success. By the nineteenth century much of the region occupied by the East Asian civilisations had also been colonised by Fertile Crescent civilisations. Only the tiny New Guinean superorganism would remain untouched, hidden in its mountain fastness, until 1935.

At the same time, profound changes were occurring in the civilisations of the Fertile Crescent. The patenting of the improved steam engine in 1775 would lay the groundwork for the exploitation of fossil fuels and thus open vast new energy stores. In the US, industrialists were perfecting means to optimise the output from their factories, introducing innovations such as interchangeable parts, the assembly line and mass production. And the nature of politics was changing too. Beginning with the establishment of the first modern republic, in the US in 1788, one monarch after another was overthrown. By 1917 unbridled monarchies had all but vanished from Europe. A system of governance that dated back to the very first cities in the Fertile Crescent had thus been vanquished, at least in Europe. From the fervour of the French Revolution to Marxism, new ideas about how society might be organised were proliferating. Together these three revolutions— scientific, industrial and political—would shape us.

By the dawn of the twentieth century, European nations had claimed most of the habitable world as colonies. In Africa and Eurasia the colonised civilisations proved more resilient than those of the New World, and European rule was often brief. The British Raj might have inserted itself at the apex of the Indian social ladder, but beneath it India continued much as it had for millennia. The same was true of most of Africa and Asia. Importantly, however, colonisation would prove to be one of the wellsprings of a migration that would in time make the capitals of the west almost as multicultural as was imperial Rome.

Throughout this period, a global human culture was also taking shape, and its source was the US. Culture often serves to set people apart from others, through diet, dress or behaviour. American popular culture has had to provide cohesion to a land of immigrants drawn from all corners of the Earth and is thus inclusive. Many aspects of it are already informing a world culture, which will doubtless evolve to vary from place to place, but will inevitably carry an indelible American birthmark.

This brief history of the human superorganism would not be complete without an account of its cumulative impact upon the planet. Professor William Ruddiman of the University of Virginia has illustrated that Earth's atmosphere provides a precise and sensitive barometer of that change. By examining the gases preserved in air bubbles in ice cores, he has demonstrated human-caused shifts in atmospheric composition go back at least eight thousand years. According to Ruddiman, CO_2 levels in the atmosphere peaked naturally ten thousand years ago at around 265 parts per million, before declining to around 260 parts per million by eight thousand years ago. But then a phenomenon occurred that has not been seen in any other ice-age cycle—CO_2 began to increase slowly so that by 1800CE it stood at 280 parts per million. This, Ruddiman postulates, was due to the clearing of forests to plant crops.[116]

Ruddiman observes a similar phenomenon with methane, whose concentration peaked around ten thousand years ago, then declined until around five thousand years ago, before increasing to its previous peak by 1800CE. This, he suggests, was due to the widespread adoption of rice-paddy agriculture, which produces much methane. Ruddiman's hypothesis goes further than this, however, seeing the human fingerprint in even the smallest variations in CO_2 concentration. A small drop in CO_2 levels between 1300 and 1400, for example, he links to the bubonic plague, which so devastated Europe's population that the forests began (temporarily) to regrow, so locking up carbon from atmospheric CO_2 in wood.

What Ruddiman's hypothesis tells us is that agriculture increased our potential to influence the dynamics of the Earth system. Hunter-gatherers could affect the atmosphere only by proxy, through exterminating mammoths, for example. But agriculture allowed our ancestors to affect the atmosphere's composition directly—by converting 'living carbon' (that is, carbon locked in living things) into dead atmospheric carbon at such a rate that the excess registered in the atmosphere. That is the hallmark of a Medean event that erodes the living fabric of Earth. The ability of our species to inflict damage began to increase, and the rate of increase began to grow.

4

TOXIC CLIMAX?

CHAPTER 12

War against Nature

Not so long ago it seemed that mankind
was a cancer on this planet.
JAMES LOVELOCK 1979

Television came to Australia in the year of my birth, 1956. I grew up with the box in the corner of the room. What I remember vividly, and what has all but disappeared from our screens today, are the serialised war dramas. Perhaps there were so many of them because they were cheap to make. After all, war matériel was everywhere, and there was no shortage of actors who knew their parts. Whatever the case, we kids were inspired by them, playing GI Joes and Japs, banging away with sticks for guns and apples for grenades, and forever asking fathers and uncles what they did 'in the war'.

Back then, with just three billion human beings, Earth seemed like a big place. Americans, Canadians and Australians revelled in a new and previously unimaginable prosperity. Soon there was a washing machine, a refrigerator, a television and a car in almost every home, and I remember clearly the joy that the arrival of each occasioned in my own household. It was an era of optimism, physically expressed in the baby boom, and of a can-do attitude epitomised by the space race, institutions like Camp Century (an enormous nuclear-powered Cold War facility dug into the Greenland icecap) and nation-building projects such as dams and highways.

For all its gloss and glory, however, this was a bleak era—one in which the bomb had decided who was fittest and provided the most conclusive proof possible of the power of reductionist science. The glowing stone and ornament of an earlier age and the exuberance of Art Deco were bulldozed and replaced by a least-cost landscape of grim functionality which sprang up row upon row in almost every city on Earth. It was an age in which ecology—destined to become the defining science of the twenty-first century—was naught but an obscure academic discipline. Defined by the Cold War, both the communist bloc and the west acted like toxic teenagers armed with guns, and were every bit as dangerous.

The nuclear era began in 1905 when a worker in a Swiss patent office discovered the relationship between energy and matter, expressing it in the eloquent equation $E = mc^2$ (E being energy, m mass, and c the speed of light in a vacuum). In short, it tells us that a great deal of energy can be derived from a small amount of matter—twenty-five billion kilowatts of energy per kilogram, to be precise. Compare this to the 8.5 kilowatts derived from burning one kilogram of coal and you can see the attraction of the atom.[117] Just forty years after Albert Einstein made his conceptual breakthrough, theory became reality when just one gram of matter was converted to energy over Hiroshima. Perhaps predictably in that tribal age, it was not a power station that our species chose to build to exploit this form of energy, but a bomb, so that 'nuclear' and 'weapon' are forever entwined in our vocabulary.

Some people seemed almost drunk on nuclear power. In 1946

the British zoologist Julian Huxley became the first director-general of UNESCO. He argued that if humans used nuclear weapons to destroy the Arctic icecap we could create both a warmer climate and new habitable lands.[118] Planning for an atomic war on nature had begun, and so widespread and effective would it become that every human being alive today carries traces of the conflict in their body.

By the 1950s both the USA and USSR had sufficient stockpiles of nuclear weapons to be able to contemplate using a few for 'peaceful purposes'. The American Richfield Oil Corporation was pondering whether nuclear power might play a role in helping to exploit Alberta's tar sands. The company executives reasoned that if it could explode a series of two-kiloton bombs below the 30,000-square-kilometre tar-sands deposit, the heat of the explosion would vitrify the sand, coating the cavity thus created with glass, while a peculiarity of the chemical structure of the tar would cause it to liquefy. When cooled, the tar would retain its more runny consistency and so fill the cavities. Three hundred billion barrels of crude oil could be made accessible by the process, the experts claimed, with no hazard from radioactivity.[119]

John Convey of the Canadian Department of Mines was a particularly enthusiastic advocate, seeing the atomic bomb as 'a new type of mining', and when the US Atomic Energy Commission conducted tests on the tar—the results of which were favourable—it looked like the world's first nuclear mining operation would proceed.[120] But then, in 1959, Canadian officials got cold feet. Politicians began worrying about how to break the news that American nuclear bombs would be detonated on Canadian soil, potentially devastating the Canadian wilderness. Planning came to a halt.

Meanwhile, far more ambitious plans for nuclear power, along the lines first suggested by Huxley, were being proposed. In 1962

Harry Wexler, chief of scientific services at the US weather bureau, fleshed out Huxley's proposal by suggesting that the weapon of choice for destroying the Arctic icecap should be the hydrogen bomb. Ten 'clean' hydrogen bombs, exploded above the icecap, would, Wexler argued, reduce the Arctic Ocean's ability to shed heat energy, and so melt the ice.[121] Other proposals for destroying the icecap included covering it with a layer of chemicals, or seeding it with algae, both of which would darken its surface and so trap heat. The Russians, who had long struggled with a frigid climate, seemed drawn to such ideas and the presidium of the USSR Academy of Sciences organised a conference in 1959 in Moscow to discuss how the ice might be melted, with two follow-up conferences held in Leningrad in the early 1960s.

The doyen of polar icecap destruction was Russian oil and gas engineer Petr Mikhailovich Borisov, who published his comprehensive plan in a small book. After long experience working in the far north he had come to believe that melting the Arctic icecap would bring untold benefits to humanity. Shipping would increase, as would rainfall in the Sahara, while the tundra would become arable, he claimed. And rising seas were nothing to worry about: the melting of the Greenland icecap, he assured us, would cause a rise of only a millimetre or two per year. At the heart of Borisov's proposal was the idea of 'giving birth to a Polar Gulf Stream'.[122] This would be accomplished via sophisticated engineering works, including the building of a huge dam across the Bering Strait. Warm water would flow into the Arctic and cold water out, the overall effect being to banish the ice. Borisov concluded that if enough water were moved in this way then Earth could even be returned to the warm state that characterised the age of the dinosaurs, when there was, he argued, little temperature difference between the equator and poles. Borisov estimated the cost of the proposal at a mere 24,000 million rubles, which could be funded by a global consortium of benefitting countries. The US's share was to be $100,000 million which, he said, 'all goes to say that the appropriations for improving the global climate are reasonable'.[123]

These insane efforts at geoengineering proved, thankfully, to be beyond humanity's grasp, but such was our love of the bomb that at least five hundred nuclear weapons were detonated in the atmosphere. The energy and radiation released frequently exceeded all estimates, and some such 'tests' were extraordinarily crude, polluting and mindless—such as the British trials carried out in central Australia in the 1950s—a matter of soaking plutonium in diesel fuel and setting it alight. And such was the overall impact of these activities that by the mid-1960s radiation had altered Earth's atmosphere in a remarkable way. There was now much more Carbon-14 in the air. The blasts released atomic particles known as neutrons, which collided with nitrogen atoms in the atmosphere, transforming them into Carbon-14, a heavy isotope of carbon (an isotope being a different type of atom of the same chemical element) with eight, rather than six, neutrons in the nucleus.

Normally, just one part per trillion of atmospheric carbon is Carbon-14, but by the 1960s the cumulative effect was so great that this figure had doubled. While Carbon-14 has now declined, evidence for its high levels can still be detected. Plants use carbon to build their tissues, and they do not discriminate between Carbon-12 (the most common isotope of carbon) and Carbon-14. Any organism that was living in the 1950s and '60s incorporated the excess Carbon-14 into its tissues. So trees, for example, which lay down growth rings, have excess Carbon-14 in rings laid down during that time.

In 2005 Swedish biologist Jonas Frisen decided that the excess Carbon-14 present in the 1950s and 60s might be useful in determining how long individual brain cells live. If the brain cells of people born in the 1950s contained Carbon-14, this was evidence that they were created around the time of birth. If, however, the cells had less Carbon-14, then they might have originated more recently, demonstrating that brain cells can regenerate.

It turned out that cells taken from the neocortex (the bits involved with memory, speech and language, among other things) were as old as the people themselves, providing proof that major portions of their brains had not produced new cells since birth. So, anyone born in the 1950s, '60s and '70s carries a permanent nuclear legacy in their brains—in the form of excess Carbon-14—which was created during the age of nuclear testing.[124] While there are no known health affects, it's a striking illustration of how pervasive the consequences of our love affair with the atom are.

Just as a doctor can use radioisotopes (atoms with an unstable nucleus, which thus produce radiation as they decay) to determine the inner workings of our body, so can scientists follow radio-active substances as they flow through the Gaian system. What they find raises the question of how close nuclear weapons have come to delivering a fatal dose of radiation to life on Earth.

We will never know with certainty, but some idea of the massive impact of nuclear testing in the 1950s and '60s was gained when researchers traced the fate of tiny, short-lived radioactive particles created during the meltdown of the Chernobyl reactor. The parti-cles were initially injected high into the stratosphere, where strong winds transported them right round the Earth in just fifteen to twenty-five days. Then, as they sank through the atmosphere, they reached the surface of the ocean. Researchers anticipated that they would stay at the surface for many months; they reasoned that the particles, being small, would take a long time to sink. Inexplicably, however, the particles vanished almost as soon as they hit the ocean. The researchers examined the entire water column in vain. It was only when they tested sea cucumbers living at depths of almost three kilometres on the ocean floor that they found what they were looking for. The particles had arrived on the ocean floor and been ingested by the sea cucumbers within a week of hitting the water.

The rapid transit of the radioactive particles was explained only when researchers began examining the role that plankton, such as krill, play. These organisms form the basis of the food chain. They spend the night feeding on tiny, single-celled plants and other creatures at the ocean surface. It's too dangerous for them to linger there in daylight, however, so towards dawn they return to the depths. The algae absorbed and concentrated the radioactive particles, and when they were consumed by the krill the radiation was concentrated again. The krill defecated only when they had reached deep, still water, and their fecal pellets, enriched with radioactive elements, dropped swiftly to the sea floor. There awaited the sea cucumbers, the ultimate garbage disposal systems in the sump of the ocean deep.

The highest level of radioactivity ever recorded in a living thing was in a krill. After the Chenobyl accident, scientists measuring radioactive fallout in the Mediterranean Sea discovered that the hepatopancreas (a sort of liver and pancreas combined) of a shrimp known as *Gennadas valens* contained Polonium-210 at a concentration of 856 picocuries per gram (a picocurie is a trillionth of a curie, which is a measure of radioactivity). Polonium-210 was used to kill the Russian dissident and counter-terrorism expert Alexander Litvinenko in 2006, and his body contained so much of it that doctors recommended that his coffin not be opened for twenty-two years. The concentration in the shrimp hepatopancreas was around a million times higher than that of the surrounding sea water.[125]

During the 1950s and '60s the oceans were our favourite dumping ground for nuclear waste. The dumping supposedly stopped in 1972, when the major industrial nations ratified the London Convention banning the disposal of highly radioactive waste at sea. But in late 2008 it was revealed that in 1978 the nuclear waste generated by British testing in Australia had been secretly dumped in the ocean, and, of course, the Russians have for years been using the Arctic Ocean as a graveyard for their superannuated nuclear submarines. Seventeen nuclear reactors that once powered submarines lie on the bottom of the Arctic Ocean, their radioactive materials just waiting for time, tide and corrosion to release them. In 1993 the London Convention was strengthened, but by then around 1.24 million curies of radiation had been dumped at seventy-three sites. The Europeans alone dumped 220,000 drums, weighing 142,000 tonnes, in the waters of the northeast Atlantic. Like the nuclear submarines, the rusting drums are an environmental time bomb.[126]

CHAPTER 13

Gaia-killers

*It is one of the great paradoxes of the human
condition that, while we are almost paranoid
in our vigilance in regard to taking toxins
into our bodies, we are all but oblivious to
the possibility of poisoning our planet.*
LINDA LEAR, INTRODUCTION, *SILENT SPRING*

Of all the human follies of the past century, few have been as threatening to life on Earth as the agricultural spraying programs that began in the 1940s. Their origins lay in chemical weapons produced by the Nazis, specifically the nerve gases synthesised by Gerhard Schrader, then working for the chemical industry conglomerate IG Farben. Schrader was put in charge of production for the Nazi chemical-warfare effort, and by the end of the war he had overseen the creation of vast stockpiles of chemical weapons so potent that even the Nazis feared using them. At the war's end, American industries gained access to these stockpiles and to Schrader's technology, and it was soon discovered that with a little tweaking even the most deadly chemicals could be put to work exterminating pests.

The commercial opportunity this presented was enormous. Not only had American companies appropriated years of Nazi-funded research, but wartime aircraft could be had cheaply, and trained pilots who wanted to continue flying were readily available. In other words, the matériel required for another war was in place. All that was needed was an enemy, and the obvious targets were humanity's oldest adversaries—the insects that bring us diseases and consume our crops. In order to make the war maximally profitable it would need to be as global as possible, and funded by government authorities. What was envisioned was a sort of Final Solution, in which chemical weapons would be sprayed across continents, transforming gardens and fields into a fertile, pest-free, weed-free paradise. What eventuated was the deaths of billions of innocent bystanders, including millions of humans, and a blighted world which even today carries a horrendous toxic legacy.

It fell to a modest marine biologist named Rachel Carson to document the unintended consequences of this mass extermination. Her book *Silent Spring* altered the course of human history, and her summary of the 'war on nature' has never been bettered:

> From small beginnings…the scope of aerial spraying has widened and its volume has increased so that it has become what a British ecologist recently called 'an amazing rain of death' upon the surface of the earth. Our attitude toward poisons has undergone a subtle change. Once they were kept in containers marked with skull and crossbones; the infrequent occasions of their use marked with utmost care that they should come in contact with the target and with nothing else. With the development of the new organic insecticides and the abundance of surplus planes after the Second World War, all this was forgotten. Although today's poisons are far more dangerous than any known before, they have amazingly become something to be showered

down indiscriminately from the skies…Not only forests and
cultivated fields are sprayed, but towns and cities as well.[127]

The substances used belong predominantly to two chemical
families: the organochlorines, of which DDT is perhaps the best
known, and the organophosphates, which include Malathion. I
use the term Gaia-killers for them, and for some other substances
discussed here, because of the way they can spread through and
destabilise ecosystems, poisoning entire food chains. Organochlorines include the 'nerve gases', which act on insects and soldiers
or civilians alike by attacking the nervous system. Because they
accumulate in the body, repeated exposures raise the risk of severe
damage to health. Around forty-two thousand cases of severe
pesticide poisoning (mostly from organochlorines) are reported
annually in the US. And we're still learning of possible links between
organochlorines and disease. For example, while not proven, there
are suggestions that certain kinds of endometriosis may result from
exposure to them.[128]

Organochlorines take a very long time to degrade, which
means that once they're out of the bottle they'll be around for
years. They are also volatile, readily entering the atmosphere, and
so spread far and wide. Because they cannot be dissolved in water
yet dissolve easily in fat, once they are taken into a living body
they tend to stay there, accumulating in fatty tissues such as the
brain and the reproductive organs. These characteristics also mean
that their concentrations increase the further up the food chain
you go. Squirrels may have a low concentration, but if a hawk eats
a hundred squirrels, it will accumulate one hundred times as much
toxin as was present in a single squirrel. Because humans are at the
top of what is often a long food chain, we're at grave risk from such
compounds. There are few ways to remove these chemicals from

our bodies, but one route emerged when researchers discovered that women generally have lower concentrations than men. This seemed like good news until they realised the toxins are present in breast milk, and that mothers were eliminating the organochlorines from their bodies by feeding them to their babies.

The nerve gases work by interfering with the way calcium is used in the body. Calcium is essential for allowing nerve cells to communicate with each other, but it is also used to build bones and the shells of birds' eggs. The organochlorine DDT had a catastrophic impact on bird reproduction: individuals affected by it produced eggshells so thin that they were crushed by the brooding mother. It was the birds' widespread failure of reproduction that alerted scientists to the Gaiacidal potential of the organochlorines.

The second family of chemical warfare agents is the organo-phosphates. Compared to the obdurate organochlorines, they have the virtue of breaking down relatively swiftly. Yet they are far more toxic than the organochlorines—minute amounts can cause death. Many have been used in warfare, of which Sarin is perhaps the most widely known. Contact with organophosphate-based insecticides remains one of the most frequent causes of poisoning worldwide, and in rural areas these chemicals are frequently drunk as a means of suicide.

A sense of what occurred during the days of aerial spraying can be gained from an incident documented in some detail by Carson.[129] She recorded that in 1959 the Detroit area was sprayed with Aldrin, the most deadly organochlorine then available and chosen simply because it was the cheapest option. The target of the campaign was a creature known as the Japanese beetle, which lives on rosebushes, crepe myrtles, grapes and a few other plants, and which was credited with minor damage to gardens and crops estimated at around $10 million per year. At the time of the program, the beetle had been in the Detroit area for thirty years, but was so rare that a well-known Michigan naturalist commented: 'I have yet to see a single Japanese beetle, other than the few caught in government catch traps in Detroit.'[130] It's hard to resist the notion that the bug's name made it an ideal target in this new war. Whatever the case, the sprayers worked with enthusiasm, spreading the toxin so that it fell from the air like snow and had to be swept off porches and paths.

Within hours of the assault the telephone of the Detroit Audubon Society was ringing hot. Mrs Ann Boyes, the society secretary, received a call from a woman who reported encountering a large number of dead and dying birds on her way home from church. Another saw no live birds at all, but found a dozen dead ones in her backyard, as well as dead squirrels. Cats and dogs sickened, and an intern at a local hospital reported treating four people in quick succession for nausea, vomiting, chills, coughing and extreme fatigue—all symptoms of organochlorine poisoning. The authorities who had ordered the spraying seemed annoyed by these complaints from old ladies and other sensitive types, the health commissioner saying that the birds must have been killed by 'some other kind of spraying', while officials reassured the population that the 'dust is harmless to humans, and will not hurt plants or pets'.[131]

To gain a sense of the extent of the attack on nature being waged back then, you have to imagine events like that being repeated tens of thousands of times, for decades, right across the US, Canada and Europe. Sometimes the war was not aimed at insect pests at all. In 1959 in southern Indiana, farmers, knowing how toxic the chemicals were to birds, deliberately sprayed the organophosphate Parathion directly onto nesting blackbirds and starlings, killing sixty-five thousand.[132] Buoyed by seemingly endless demand, the scale of production of such chemicals grew. By 1960 around

290 million kilograms, made up of about two hundred different chemical compounds, were coming out of the chemical factories of the US per year, and they were being applied 'almost universally to farms, gardens, forests and homes'. According to one expert quoted by Carson, the amount of Parathion being used each year in California alone at the time was enough to kill the human population of the world five to ten times over.[133]

Because the toxins spread swiftly and are cumulative, not a single ecosystem escaped their influence, with waterways being particularly vulnerable. Streams in more industrialised areas were being entirely emptied of fish and other life. This was the age of the death of the Thames, the Yarra, the Cuyahoga and many other rivers that were turned into industrial sewers because they had the misfortune to flow through a centre of industrialised 'civilisation'. Even relatively pristine rivers such as New Brunswick's Miramichi did not escape. In the late 1950s spruce budworm was detected in the forests of the region, and spraying for them devastated the region's salmon run. Yet neither the government nor the corporations involved showed the least concern. They kept issuing mollifying statements, dismissing reports of damage as the complaints of nature lovers who were against progress.

A few dedicated naturalists, however, were collecting evidence of the long-term consequences of insecticide use. Charles Broley was a retired banker from Winnipeg. Whatever he achieved in his financial career has been long forgotten, but his passionate pursuit of a hobby has given him a place in history. Broley was fascinated with America's most majestic bird, the American bald eagle, and he had been banding young eagles at their nests in Florida for a decade before the insecticides became commonly used. He recorded that before 1947 he would band an average of 150 young eagles from 125 nests each year. By 1958, however, he had to search 160 kilometres of coast before banding the single eaglet he found that year. Over the intervening decade, the use of organochlorines had become widespread, and because eagles sit at the apex of the food chain they had become the ultimate repository for the chemicals. The sick birds suffered severely impaired fertility, and if they did manage to conceive they laid eggs with shells so thin that they would crack before the chicks hatched. Broley had caught the species in the act of vanishing from the lower forty-eight states of America. Tragically he would not live to see its return as he died in 1959. Just a decade later the bald eagle reached its nadir and people feared it might vanish entirely from the region.[134] Thanks to the Broleys and Carsons of this world, however, it was given a second chance. Following the banning of DDT in 1973 it began to recover. On 28 June 2007 the US Interior Department removed it from the endangered-species list, and it became one of just a handful of species ever to fight its way back from near extinction. Every time I see one of these magnificent birds perched by a river or soaring overhead, I feel hope for a better world.

Perhaps the most astonishing aspect of this war on nature was its manifest failure to achieve its goal, for the pests were becoming more abundant than ever. Certainly their numbers decreased immediately after spraying, but the spray killed their predators too, and because the pests were small and reproduced rapidly, their numbers rebounded long before their predators' did. Even worse, they rapidly gained immunity to the sprays. As early as 1959, over one hundred major insect pest species were showing signs of resistance to the toxins. And humans responded with repeated sprayings of the chemicals at higher concentrations, which of course only worsened the problem.

Long before the development of organochlorines, intelligent farmers and entomologists knew what the solution to pest infestations was. Where pests had been introduced from elsewhere, it often paid to introduce their predators as well. Increasing ecosystem health by rotating crops, providing companion plantings that aid predators or deter pests, and a tolerance of low levels of pests were all part of a tried-and-true method of control that long predated the war on nature. The trouble was that corporations couldn't make money from these approaches. With the illusion of a quick and permanent fix, the pesticide companies had set us on a cataclysmic course.

Silent Spring appeared in 1962, but it took until 1973 for the US to ban manufacture of some of the most dangerous organochlorines. Today organochlorines, along with organophosphates, continue to be used in many developing countries. While there is some justification for restricted use of such chemicals, the situation in many instances seems to be as bad if not worse in the developing world today than it was in 1950s America. In 2007 the annual global human death toll from insecticide poisoning was approximately 220,000, with around three million suffering from severe but non-fatal exposure, mostly in the developing world.[135] Those in the west are not insulated from the misuse of these chemicals. Just look at the labels indicating the origin of the fruit and vegetables in your supermarket. The truth is that, for all of our domestic precautions, we are all inextricably connected through webs of food, other traded commodities and our atmosphere. Whatever misfortune befalls one part of the Earth will inevitably be widely felt.

There is another reason, however, that we should not feel safe from the toxic 1950s and '60s. Many of the chemicals sprayed back then have been faithfully stored in our bodies all those years, and are immoveable until the moment of our death. Even if you were born in later decades, you're still likely to contain some of them, having imbibed them with your mother's milk. The high rates of some cancers in rural communities may have their origins in the sprayings of fifty years ago, as may decreases in male fertility such as that documented in Denmark, which has kept accurate figures in this obscure area of human health.

In Denmark in 1940, the average human ejaculation consisted of 3.4 millilitres of semen with 113 million sperm per millilitre. By 1990, however, the volume had declined to just 2.75 millilitre and the sperm concentration to 66 million per millilitre. Such figures could result from Danes having more sex in recent years, but because these reductions are accompanied by other changes it seems more likely to be a genuine environmentally caused decline. The rate of testicular cancer has increased in Denmark to five times the rate in Finland, which tellingly has less agriculture.[136] And this is probably not a uniquely Danish problem—some researchers argue that similar declines have occurred in many countries but have gone undetected.

Organochlorines and organophosphates are not the only Gaia-killers. From the 1940s to the 1970s polychlorinated biphenyls (PCBs) were used extensively in manufacturing. Today, they mostly make the headlines when old industrial sites are being redeveloped, as they are extremely persistent and can remain in the soil for decades, centuries or perhaps longer. Use of these toxic chemicals was particularly prevalent in the electronics industry, but they also found uses as coolants and lubricants, pesticide extenders and in the manufacture of carbonless copy paper.

It was recognised from the very beginning that PCBs had all the hallmarks of dangerously toxic compounds. They are soluble in fats and oils, are readily absorbed directly through the skin and are extremely stable—attempts to break them down often result in the production of even more toxic daughter products. As early as 1937 the Harvard School of Public Health hosted a conference on the dangers, enough being known at that time for a leading chemistry journal to publish an article that called them 'objectionably toxic'.[137] Despite the warnings PCB use continued unrestricted until the US Congress banned domestic production in 1977. Today we know that PCBs are virulent carcinogens; they also cause liver damage, result in poor cognitive development in children and engender reproductive disorders and birth defects, including poorly developed sex organs and hermaphroditism.

What makes PCBs almost archetypical Gaia-killers is their capacity to evaporate readily, thereby entering the atmosphere. Once airborne they are swiftly transported globally. Since oceans cover 70 per cent of our planet, most PCBs end up falling into water. In 1971, a US National Academy of Sciences report found that within just a year of their manufacture one quarter of PCBs would end up in the oceans. Just what they did there was discovered only accidentally, when scientists studying phytoplankton noticed abnormal reproduction and metabolism of the samples in their bottles. Eventually they realised that the latex seals on the bottles were releasing almost undetectable amounts of PCBs into the ten-litre water samples they held. Subsequent laboratory experiments have shown that as little as one part per billion depresses reproduction of marine algae and other marine plants by half, and there are sufficient concentrations of the toxin in wild-living algae worldwide that experts say they 'believe that such levels must be affecting the marine biota'.[138]

Unfortunately the full impacts of PCBs on the oceans will never be known, for they became widespread long before baseline studies revealing how things were before they arrived could be carried out. Evidence is emerging, however, of serious impacts even in remote oceans: polar bears and whales have been suffering the same kinds of birth defects that PCBs induce in humans, including hermaphroditism and other genital deformities. The least known yet potentially most devastating impact of PCBs is probably occurring in the ocean depths. Because PCBs concentrate in algae, and become further concentrated in krill, the levels reached in krill faeces can be 1.5 million times higher than in surrounding sea water. Krill faeces sink to the bottom of the ocean swiftly, and so little sediment falls into the ocean depths that they are buried at the rate of just one millimetre per thousand years. This means that the chemicals remain available to living things for thousands of years.[139]

More chemicals in widespread use are identified as potentially toxic every year. From polyfluoroalkyls to bisphenol A, from hexa-bromocyclododecane to dibromopropyl phosphate, our domestic environments are saturated with compounds that are now known to have adverse affects. We are without doubt the most toxin-drenched generation in human history and, as dangerous as any individual toxin is, it's the total burden of toxins in our environment that counts. No group is so vulnerable to this rain of chemicals as is the young. Their metabolisms run at breakneck speed as they grow, allowing free rein for toxins to cause damage. In 2000, the US National Academy of Sciences estimated that fully one-quarter of all developmental problems seen in children 'were caused by environmental factors working in combination with a genetic disposition', citing the increase in infants born underweight or prematurely, heart defects, genital defects and nervous-system disorders.[140]

In light of this sorry history, it would be easy to blame the Medean nature of the post-war period on unconstrained human ingenuity, or a rampant capitalist system. But evidence from the communist countries suggests that something far deeper was at work, for those countries mounted their own wars on nature which were, despite the lack of chemical weapons, as lethal as those of the west. In Mao's China, brute human effort was the tool of choice which, following the aphorism *ren ding sheng tian* ('man must conquer nature'), in just a few decades, turned China into an environmental basket case.[141]

Mao's war on nature reached a peak around 1958 to 1960. In a famous and widely reported campaign the population was mobilised to bang pots and pans together until the sparrows and other birds fell to earth, dying of exhaustion. The idea was to spare grain crops from their depredations. But because insect-eating birds were persecuted as vigorously as the grain-eaters, the result was to leave the fields vulnerable to insect attack. The mass relocation of population during the Maoist era led to further devastation. The rainforests of southern China, which had been previously spared because of steep slopes and poor soils, were settled and felled, and the land left in ruin. Delicate grassland ecosystems in the west were put under the plough, with equally appalling results, while elsewhere forests were felled to fuel useless backyard iron smelters. Many of these vulnerable ecosystems have never recovered, and today their destruction continues to engender landslides and dust storms. To quote one expert, 'Air and water pollution...deforestation, erosion, desertification, habitat destruction, and falling water tables followed in the wake of [the] program.'[142]

But perhaps the most enduring environmental legacy was Mao's encouragement of population growth. At the beginning of his time in power China had around five hundred million people, and was widely believed by experts to be on the verge of dangerous overpopulation. Yet Mao made contraceptives hard to get and exhorted women to have large families. As one Chinese census official said, 'Chairman Mao has direct responsibility for the population problem. As Party Chairman, policy depended upon him. When he said, "With Many People, Strength is Great" that was the last word on the subject.'[143]

A key lesson of the toxic 1950s and '60s is that it's our beliefs about our relationship to each other and to the world, rather than to our technology, that determine whether we show a Medean or a Gaian face. Another lesson is that from the remote poles to the ocean depths, and from our brains to our gonads, our actions have marked life indelibly. It's a taint that will be readable down the ages, and it will signify forever a time when humanity, with an appalling lack of wisdom, decided to wage war on nature. That war has continued almost to the present; now, at the eleventh hour, hope of decisive change has arisen.

CHAPTER 14

The Eleventh Hour?

*One can see from space how the human race
has changed the Earth…human exploitation
of the planet is reaching a critical limit.*
STEPHEN HAWKING 2007

In 1989, the communist bloc of nations began to collapse, destroying the Cold War paradigm of a world divided into two huge alliances eternally at each other's throat. Within a few years the internet was being used by hundreds of millions, and we found that we could communicate in an instant with people almost anywhere. At the same time, economies across the planet were becoming more interdependent and intermeshed. All of these changes made it increasingly evident that nations acting alone were simply not powerful enough to overcome our greatest challenges, and international treaties and agreements capable of securing a safe environment, peace and human dignity were being pursued. By 1996, seventy-one states (including five of the eight then nuclear-capable states) had signed the Comprehensive Nuclear-Test-Ban Treaty, which bans the testing of nuclear weapons in all environments. By early 2009 it had been signed by 180 heads of state and ratified by 148 of their governments, though the US, China, India, Pakistan, Israel and North Korea—all of which have nuclear weapons capacity—are not among them; nor is Egypt, Indonesia or Iran. In May 2001, ninety-one countries (including the US) joined together to ratify another crucial treaty—this one outlawing some of the weapons used in the war on nature. Called the Stockholm Convention on Persistent Organic Pollutants (POPs), it grew out of discussions at the 1992 Rio Earth Summit, which at that time was the largest global environmental meeting ever held.

While the convention on POPs has received little publicity, it may well prove to be one of the most important initiatives of the early twenty-first century. The chemicals banned under it possess characteristics that make them uniquely dangerous to life: all are toxic, extremely resistant to degradation, and able to spread swiftly in air and water, to accumulate in body fats and oils, and to be passed on from mother to young. Such chemicals are impossible to contain; the damage they do commences from the moment of their manufacture. The 'dirty dozen' recognised under the treaty are: the insecticides Aldrin (first manufactured in 1949), Chlordane (1945), DDT (1942), Dieldrin (1948), Heptachlor (1948), Mirex (1959) and Toxaphene (1948); the rodenticide and insecticide Endrin (1951); the fungicide Hexachlorobenzene (1945); PCBs, and the by-products Dioxins and Furans. The POPs treaty came into effect in May 2004 and is strongly worded, demanding the outright banning and destruction of the POPs chemicals. It also seeks to prevent the emergence of new dangerous chemicals, and commits developed countries to assist developing nations financially to achieve these goals.[144] But can we be sure that all of the Gaia-killers have been identified and banned?

One class of chemicals the treaty signatories are keeping an eye on but have not yet banned is polyfluoroalkyl substances, or PFSs.

PFSs are used in lubricants, adhesives, and soil and stain repellants, as a paper coating and in personal care products. They degrade into persistent chemicals known as PFCAs and PFSAs, which can accumulate and have been detected in all kinds of vertebrates. They are pervasive in human tissues. People living in industrialised regions carry particularly high concentrations. There was a marked increase in these concentrations during the 1990s, as new uses were found for the chemicals.

The means by which these chemicals find their way into our bodies is somewhat unusual. The key source is probably personal care products, along with indoor dust. Children have particularly high concentrations, perhaps as a result of playing on carpets and furniture treated with stain protectors. PFSs have been detected in menstrual blood, human milk and semen, indicating that they are present in the reproductive system and are transferred to infants.[145] The precise effects of these chemicals on the human body are still being ascertained, but they've recently been linked to low birth weight and small newborn head size.[146] In laboratory animals concentrations lower than those often observed in humans have caused cancers and poisonings. Earlier indications of toxicity induced the principal manufacturer in the US (the 3M Company) to phase them out, and by 2004 global production was down to half that of four years earlier.

Conventions such as the POPs treaty cannot provide the complete answer to our chemically caused problems, because several classes of toxins cannot be regulated in such ways. An intriguing example concerns substances that do us good, but which are fatal, even in minute concentrations, to other species. The first years of the twenty-first century were heralded in India with the mass deaths of vultures. Though hardly creatures to evoke sympathy, vultures play a critical role in ecosystems through the swift recycling of carrion. Three species—the slender-billed (*Gyps tenuirostris*), the Indian (*G. indicus*) and the white-rumped (*G. bengalensis*) once abounded on the subcontinent. Indeed, until two decades ago the white-rumped vulture was the most abundant large bird of prey on Earth. Yet today it's one of the rarest, and listed as critically endangered. Until 2006 nobody had the slightest idea what was causing its decline. Neither did many care.

Beginning in East Asia in the late 1990s, the vulture decline swiftly moved west into the Indian subcontinent, where it caused 'such a large-scale die off of this hardy group of scavenger birds [as] is unprecedented in the world'. In fact the decline was even faster than that of the dodo as it slid towards extinction in the late seventeenth century. In just fifteen years the vulture population plummeted by 97 per cent. Inevitably, the number of unscavenged cattle carcasses grew until they became a health hazard, while the number of feral dogs increased, raising fears of a rabies epidemic. Despite the scale of the ecological disaster, few Indian bureaucrats seemed to be concerned. Indeed the Indian Ministry of Environment and Forests initially refused permits to researchers to capture dying birds so that the cause of their deaths could be ascertained.[147]

There was one group of Indians, however, that was deeply traumatised by the catastrophe. The Parsees practise one of the most ecologically sustainable religions on Earth, and it is their custom to expose their dead on towers, known as the Towers of Silence, where vultures consume the bodies. This mortuary practice dates back to the sixteenth century, and is preferred by Parsees because they believe that the five natural elements—water, soil, fire, air and sky—are sacred and should not be polluted. They believe that, after the soul leaves a body, the corpse should be disposed of with minimum harm to other living things. By transforming the dead into living flesh again, vultures complete the life cycle in an environmentally sound manner. But as the vultures vanished, the bodies remained uneaten, and so began to decompose and smell, polluting the air and water. 'There are horror stories of bodies piling up over weeks and months as vultures have stopped visiting the Towers of Silence,' one anxious official of the Bombay Municipal Corporation said.[148] Desperate, the Parsees experimented with solar mirrors, hoping their heat would help complete the rites, but with little success.

Many believed that the scavengers were being laid low by a virus or some other pathogen, yet no sign of a disease-causing organism could be found. Autopsies revealed, however, that the vultures had suffered acute kidney failure, which is an adverse side effect of some medicines. It turned out that a cheap new anti-inflammatory drug, Diclofenac, had become available, and was being widely administered to cattle and water buffalo. Anyone who has suffered from inflammation or gout will know Diclofenac, perhaps under the brand name Voltaren. It is swiftly effective to ailing beast and human alike, but uniquely lethal to vultures. And of course some of the cattle treated with it failed to respond and died loaded with the medicine. Studies revealed that if just one dead cow in a hundred had Diclofenac in its body the vultures would be driven to extinction. In fact around 10 per cent of cattle carcasses sampled in India at the height of the die-off contained the drug.

In 2006 the Indian government banned Diclofenac for veterinary use, and pharmaceutical firms were told to promote an alternative known as Meloxicam instead, which has no adverse effect on vultures.[149] Diclofenac, however, is still widely used by humans, and there's little to stop farmers giving it to cattle. The Indian environmental group Centre for Environment Education is trying to prevent this by giving partial rebates to farmers using Meloxicam for their cattle, and together with government regulation this is slowly having an impact. But India's vultures were brought so close to extinction that it will take decades for them to recover. Still, in April 2008 a sight was seen in Mumbai (where most Parsees live) that gladdened many hearts. After a complete absence of more than three years, a flock of around twenty vultures was sighted near the Alipore zoo and racecourse. Whether their numbers will grow to the point of reviving the Towers of Silence remains to be seen.

Many animal species suffer mysterious declines, but unless the creatures are directly relevant to humans and our economy—such as honeybees, which recently suffered a still unexplained population collapse—we rarely hear about them. Not all of course may be caused by human activity, but we'd be wise to keep our minds open to the possibility. One of the most catastrophic declines occurring right now is that of Earth's amphibians. Fully one-third of all species of frogs and toads have declined in recent years, and a large number of species have become extinct. These include beautiful creatures such as Costa Rica's golden toad, and Australia's two species of gastric brooding frogs, which incubated their eggs in their stomachs. Among the culprits identified are climate change and the global spread of the chytrid fungus, which attacks the parts of the frogs' skin that makes keratin.

Genetic studies indicate that chytrid fungus originated in southern Africa, and that it may have spread via African clawed toads. These flattened, mottled brown creatures, which are about twelve centimetres long, are notable for lacking tongues and external ears—handicaps which, along with their enormous hind legs and ridiculously small heads, make them appear to be the most inept of animals. But with our help they have achieved a global distribution, at least in medical laboratories, and for the strangest of reasons.

Prior to the development of home-test kits, they represented the best method of detecting pregnancy. A technician would inject a woman's urine into the toad, and if it laid eggs within twelve hours it was a reliable sign that the woman was pregnant. Unfortunately, control of the toads was not always strict. A doctor friend told me that when he worked in Papua New Guinea he'd seen the creatures hopping around the pathology laboratory, the women's name tags attached to their legs! Whether any slipped through a door into the tropical night remains an unanswered question.

The chytrid fungus first showed up among wild-living amphibians in 1988, around the time that the toads were being replaced by kits. Perhaps some well-meaning lab technician decided to release his toads rather than euthanase them, and so unleashed this grim pestilence on the amphibian world—or perhaps some had escaped earlier. Whatever the cause, the result has been catastrophic. And if such things can happen on land and go undetected for years, imagine what could be happening in the sump of the Earth—the ocean abyss, the largest habitat on Earth, where many of our most toxic chemicals accumulate. Marine biologists have warned that the ocean depths 'may be the first global biotic environment that faces a long-term danger from contamination'.[150] Yet such is our limited knowledge of our living planet that if the deep was already in the throes of such a crisis we simply would not know.

The final kind of manufactured environmental toxins we must consider poses no direct harm to any individual being, but nevertheless threatens all life on Earth. By far the most important are the chlorinated fluorocarbons (CFCs). First manufactured in Germany in 1928, they soon proved to be enormously useful, and by the 1970s they were in a range of goods, including polystyrene cups, refrigerators and propellants in aerosol sprays. Moreover, they were long considered environmentally friendly because they did not interact with anything at the surface of the Earth—and if they did not react with any living thing, how could they harm it? Yet harm was indeed being done.

Commencing in the International Geophysical Year of 1957, measurements of the atmosphere above Antarctica began to be taken, and by the 1970s they showed a clear seasonal decline in ozone concentration in the stratosphere. Ozone is a molecule made up of three oxygen atoms, and in its pure form is a beautiful, duck-egg-blue gas. Just three out of every ten million molecules in the atmosphere are ozone, and they are concentrated twenty-five kilometres above our heads. Ozone is vitally important to life because it blocks nearly 99 per cent of the ultraviolet radiation headed to Earth. It is Earth's sunscreen, and is as much a product of life as the more common form of oxygen, O_2.

Years of scientific research finally revealed CFCs as the cause of ozone decline. This was astonishing to researchers because there are just a few parts per billion of them in the air—so few indeed that they could not be detected until James Lovelock devised a special machine to do so. As CFCs drift up into the stratosphere, ultraviolet radiation breaks them down, releasing atoms of chlorine. A single free chlorine atom can destroy a hundred thousand ozone molecules before it is locked away into some other molecule.

News of the decline in ozone triggered widespread alarm. Ultraviolet radiation is extremely dangerous because it penetrates cells, tearing apart DNA and disrupting metabolic processes. Ozone protects us from much but not all of that damage. Lie without sunscreen on a tropical beach and in twenty minutes you're likely to have a nasty sunburn. Without ozone, you'd get the same burn in a matter of seconds.

The prospect of an Earth without ozone protection was so alarming that in September 1987 the nations of the world came

together in Montreal to begin the process of banning the production of CFCs. The treaty was swiftly and effectively executed, an
action to which we owe the present quality of our lives. We know
how much of the chemical was being produced, and can determine
what the impact on the ozone layer would have been if the treaty
had failed: by 2008 there would have been a large and permanent
ozone hole over the South Pole, a second permanent hole over
the North Pole, and dangerously depleted ozone down to middle
latitudes elsewhere. This would have led to a world crisis, as with
every 1 per cent increase in ultraviolet radiation there's a 1 per cent
decline in seed germination, a 1 per cent increase in blindness, an
increase in skin cancer and metabolic disorders, and a decline in
our immune systems and in oceanic productivity.

Despite the ban, by the mid-1990s the hole in the ozone layer
over the South Pole had grown to cover an area as large as North
America, and a second, smaller hole had appeared over the North
Pole. This occurred because CFCs last around fifty years in the
atmosphere and take five years to reach the ozone layer, so for a
while after their prohibition an ever-larger plume of CFCs was
reaching the ozone layer. But now the ozone hole is declining, and
perhaps one day will vanish. Then, I hope, humanity will proclaim
its first truly global day of celebration—16 September, for on that
day in 1987 the Montreal Protocol received its first signatories, and
a door to salvation for all humanity was opened.

CHAPTER 15

Undoing the Work of Ages

Some of them gets lead-pisoned soon,
and some of them gets lead-pisoned later,
and some, but not many, niver; and 'tis
all according to the constitooshun sur.

An Irishwoman's reflection
to Charles Dickens
in *The Uncommercial Traveller* 1876

One human activity threatens our future more immediately than any other. Mining conducted without regard for the Gaian system may seem like a harmless activity, but as we commandeer resources from wherever we find them—be it the depths of the ocean, the icy poles or kilometres deep in the Earth's crust—we start to undo the elemental concentrations that life has created between Gaia's three organs. Elements have been moving in air, land and water since time immemorial, but today we are digging them up at an unprecedented rate and scale.

The nuns who taught my wife in the 1970s possessed astute environmental awareness, and one of their lessons left a deep impression on her. The nuns told of a terrible tragedy in Minamata Bay in Japan, which occurred shortly after the atomic bomb was dropped on Hiroshima. What my wife remembered most vividly was mention of the bomb and the fate of the area's cats, which had began to walk backwards, their brains having been destroyed by a terrible toxin. Her recollection is wonderfully illustrative of the way children (and some adults) see the world—the danger of atomic weapons (so immediate and horrific) replacing the real culprit, which is invisible and slow-acting. And the focus on cats rather than people! But it is also the story of everyman, for as we seek explanations for events, we are prone to misremember and to jump to the wrong conclusions.

In reality the atomic bomb had nothing to do with Minamata's backwards-walking cats. Instead, the unusual feline behaviour was caused by mercury poisoning. Minamata Bay was being used as a dumping ground for waste mercury by the Japanese Chisso Corporation, and the cats fed on the local fish. Beginning as a fertiliser manfacturer, Chisso developed into a petrochemical producer, then a plastics company. Between 1932 and 1968 it dumped twenty-seven tonnes of mercury into the bay. Ironically, it was the villagers of Minamata who in 1907 had persuaded the founders of the Chisso Corporation to establish their factory in the area. They were poor fishermen, and the prospect of employment was a great attraction. Soon, however, pollution started to affect the fishery, and rather than clean up production the company paid off the locals to silence their complaints.[151]

Fish are an important part of the diet of the people of Minamata. A virtual pollution highway was created between the factory and the villagers' bodies, and by the 1950s people were reporting numbness in their limbs, slurred speech and problems with their eyesight. A few began to shout uncontrollably and were thought to have gone crazy. The villagers also noticed what they interpreted to be 'cat suicides', and birds falling dead from the sky. By 1959, doctors working for Chisso had established that mercury pollution from the factory was the cause, one even demonstrating its effect on cats to company managers. The company responded with years of deceit, cover-ups and bullying. Only in 1968 did it stop polluting, but by then over ten thousand people, many of them children, had suffered severe, irreversible physical and mental damage.[152]

Humans have been using mercury for at least six thousand years. In Roman times it was used to recover gold from ore (a use which continues today), and over the centuries it has been used to make a wide variety of products, from medicines to hats, along the way killing many patients and making many hatters mad. Today mercury continues to be used in an enormous variety of products, from fluorescent lights to dental fillings and batteries. But these sources account for only a small proportion of the mercury in the environment. Astonishingly, around two-thirds of the mercury circulating in Earth's air and oceans was not deliberately mined, but comes from the burning of fossil fuels, mostly coal.[153] Some coal is rich in mercury because coal is a natural sponge that absorbs many substances dissolved in groundwater, from uranium to cadmium and mercury. When the coal is burned to provide steam for electricity production, these elements are released into the atmosphere, then blown by winds over the ocean, into which they eventually fall.

In order to prevent the spread of mercury pollution, some governments have insisted that it be captured at the power plant smokestack, and to this end filters of activated charcoal are used. But while the mercury is captured quite effectively, no safe way has yet been found to dispose of the contaminated charcoal, prodigious amounts of which are now accumulating in warehouses and deep mines. And they are all just a fire away from releasing the mercury into the atmosphere.

Abandoned coalmines are another source of mercury. Groundwater often accumulates in the mines, and then finds its way into creeks and rivers. One example is the Canyon colliery in Australia's Blue Mountains. In 2009, a decade after its abandonment, it was found to be leaching heavy metals—including zinc at concentrations over two hundred times the safe limit for marine life. As a result fish had been eliminated for kilometres downsteam, and other life was severely affected.[154]

Mercury can enter the atmosphere from other sources. Dentists have been using an amalgam that contains mercury to fill cavities in our teeth for decades. In its elemental form the mercury in the amalgam is relatively harmless, even if, like me, you have a head-full of filled teeth. But when we die, if we choose to be cremated, the mercury in the amalgam is released, and it joins the plume of mercury issuing from the smokestacks of coal-fired power plants, rising high into the atmosphere and likely to fall into some distant sea.

Mercury remains in its elemental form while at the sunlit surface of the ocean. But, aided by the same plankton and krill that absorb radioactive particles, it soon reaches the abyssal depths, and there it is transformed into a highly toxic form known as methyl mercury.[155] No one is entirely certain how this transformation is accomplished, but it's likely that bacteria play a significant role. Methyl mercury is dangerous because it is readily taken into living organisms, and in marine creatures such as fish, shellfish and shrimps it bio-accumulates—it enters their tissues and stays there. If krill eat the bacteria containing it, they will receive a certain dose. But if a small fish eats a hundred krill, it will take in a hundred times that dose. And so it goes on up the marine food chain, until the largest predators such as sharks and swordfish accumulate dangerous levels of mercury. And what eats sharks and swordfish? The world's ultimate top predator and repository for anything that bio-accumulates: humans.

Not all methyl mercury is created in the deep ocean. Some forms in deep lakes and in landfills, wherein lie countless mercury-containing discarded batteries, fluorescent lights and other products. Nevertheless, it is a remarkable fact that much methyl mercury has taken a journey right through the three organs of Gaia—from its source in a coalmine, to the atmosphere, and on to the ocean depths—before it comes up again through the food chain and lands on our plate. It's a perfect example of how everything is connected on planet Earth.

Mercury levels in fish have tripled since the pre-industrial era, and fish exposed to non-lethal amounts of methyl mercury have such damaged nervous systems that they have difficulty escaping predators, meaning that the mercury they contain is even more likely to migrate up the food chain.[156] Many commonly eaten fish (especially larger individuals from regions such as the Mediterranean) exceed US and EU health standards. And for larger, predatory species such as swordfish, the amounts they contain can severely damage their reproductive systems. But mercury is most dangerous when it enters humans. In the US, mercury levels are four times higher in fish-eaters (defined as those who have eaten three or more servings in the past thirty days) than others, and high levels of methyl mercury can cause myriad symptoms in humans, from sweating to sensory disturbance, and damage to nerves, kidneys, liver and testes. Mercury is particularly dangerous to foetuses, hence the health warnings about fish consumption issued to pregnant women. The scale of the mercury problem in the US has only recently become evident. By one estimate, one in twelve women of child-bearing age has a blood mercury level in excess of that considered safe by the US Environmental Protection Agency (EPA). There may be as many as 4.7 million women with elevated levels of mercury, which puts 322,000 newborns at risk of mercury-related brain damage each year.[157] Under the Obama administration, the EPA is moving to address this threat, and has said it will propose standards for coal-fired power plants by 2011. And there is now global action, too, with the 2009 US endorsement of negotiations to finalise a global treaty on mercury.

Hatters were driven mad by mercury, which they used to remove the hairs from animal skins. Today, courtesy of mining, it seems that the entire world is in danger of becoming as mad as a hatter. Of course, there are solutions to the problem. If we stopped burning coal the lion's share would vanish immediately. If we sorted rubbish to remove mercury-containing waste we could isolate another important source. And we could eliminate mercury from cremations with the cheapest of solutions—a pair of pliers. But such things will not happen of their own accord. We need to create incentives and disincentives that will see the swift prevention of all major sources of mercury emissions.

Some toxic metals are useful in small amounts—the poison is in the size of the dose. Others are of no use to us, but accumulate in our bodies because they are mistaken for elements that are required as catalysts for important reactions in the body. Often the toxic metal binds more strongly to our tissues than does the useful one, making it difficult for us to utilise the beneficial element, and impossible to get rid of the toxin. Cadmium, for example, is so similar to zinc that it is carried into our bodies by zinc-binding proteins. But cadmium binds itself into our chemistry ten times more firmly than zinc, making it extraordinarily difficult to remove. And, because it replaces zinc, cadmium can interfere with the uptake of iron and calcium, which require zinc, leading to deficiencies that result in anaemia and bone disorders.

While most cadmium enters our bodies via our stomachs, it is most efficiently absorbed via the lungs. Tobacco plants accumulate cadmium in their leaves, which means that smoking is a major cause of high cadmium levels.[158] Anyone unfortunate enough to absorb a fatal dose of cadmium (just 250 milligrams can kill you in ten minutes) may well, if they survive the initial poisoning, succumb to kidney failure. The consequences of continuous ingestion of smaller amounts, however, are much more varied, and include softening of bones (sometimes to the point where they fracture merely from the weight of the body), lung diseases and possibly cancer.

Death from chronic cadmium poisoning is extraordinarily painful. A tragedy occurred in Toyama prefecture in Japan in the first half of the twentieth century. The source of the cadmium was poorly controlled mining, and those who suffered from the poisoning called it the *itai itai* (meaning 'ouch, ouch'). Most prevalent among women who had passed menopause (and therefore were already at risk of osteoporosis), it caused excruciating pains in the limb bones and spine, which became ever more brittle. The women were also highly anaemic, coughed a lot (due to damage in their lungs) and suffered from kidney failure. Despite its horrendous symptoms, the poisoning was little investigated or understood, and until 1946 it was thought to be a regional illness or specific bacterial infection. Not until the late 1960s did the Japanese government publicise the link with cadmium and poor mining practices.

Lead is similar to cadmium in the way it poisons us. It too can mimic calcium, zinc and iron, and, while it plays a useful role in our body (for example, in helping the production of red and white blood cells), when we get too much it prevents other beneficial elements from doing their jobs. If concentrations are high enough, it can inhibit vital enzymic reactions, so causing severe damage or death. Lead poisoning has a particular horror in that it can profoundly affect the growing brain. Two remarkable albeit

controversial studies following the ramifications of lead poisoning in children make clear some of the possibilities. A 2007 study
across nine countries revealed a 'very strong' correlation between
high lead levels in preschool children and subsequent crime rates,
with murder showing a particularly strong correlation with the
more severe cases of childhood lead poisoning.[159] In a second study,
Ezra Susser and colleagues from Columbia University studied
twelve thousand children born in Oakland, California, between
1959 and 1966, whose mothers had given blood samples while they
were pregnant. It was found that children exposed to high levels of
lead in the womb were more than twice as likely to become
schizophrenic.[160]

Lead has been banned from petrol, paints and other products,
but lead poisoning still occurs. The way lead gets into the foetus
is insidious. Most of the lead that makes its way into our bodies
is incorporated into our bones, and a woman damaged by lead
poisoning as a toddler will store the toxin in her body until she
herself becomes pregnant. Then, she draws on her bone reserves
of calcium and phosphorus in order to grow her baby's skeleton,
and in the process the lead is released into her bloodstream and
incorporated into the foetus.

Lead, copper, silver and zinc have been mined and processed at
Mount Isa in Queensland's Gulf Country for decades. The process
results in large emissions of toxic elements: in 2005–06 alone, an
estimated 400,000 kilograms of lead, 470,000 kilograms of copper,
4800 kilograms of cadmium and 520,000 kilograms of zinc were
released into the atmosphere.[161] Blood tests carried out in US
laboratories on six-year-old Stella Hare, who lived in Mount Isa,
revealed dangerously high levels of lead and ten other metals in
her blood. She suffers learning and behavioural difficulties, and
her family is now suing mining company Xstrata. Stella's case is

not isolated: of four hundred children whose blood was tested by Queensland Health, forty-five had lead levels above the dangerous threshold of ten micrograms, with nine above fifteen, two above twenty and one registering 31.5.[162] That's more than 10 per cent of all children sampled having dangerous levels of lead poisoning, drawn from a population which, by virtue of the boom-and-bust nature of mining, can be assumed to be reasonably transient. Some in the company deny that the lead in the children comes from the mine, arguing instead that it comes from natural sources.[163] Lead levels in soil in residential areas of Mount Isa are thirty-three times higher than federal limits, while lead in swimming holes near the town exceeds those limits several hundred times.[164] In a recent article researchers have argued that industry and government authorities have downplayed the risk of ongoing, chronic low-level lead exposure.[165] In a newspaper advertisement Xstrata claimed that 'a few simple steps related to hygiene and nutrition will ensure…lead levels remain below World Health Organization standards'.[166]

The list of toxic metals goes on and on. Arsenic insinuates its way into our bodies because it is very similar to phosphorus, while lithium mimics potassium, calcium and sodium. Lithium can be important in medicine (in treating various psychiatric disorders such as mania and depression), but the difference between a medicinal dose and a toxic one is small. Arsenic has also been used as a medicine (indeed it's currently being investigated to treat some rare leukaemias). And it was popular in the eigthteenth century as a stimulant. Too much, however, and the effects can be catastrophic. The region around Zloty Stok in Poland has long been mined for gold and silver. Arsenic was used in the mining process and it leached into local waterways. For centuries inhabitants of the town, then called Reichenstein, suffered from 'Reichenstein disease', which was characterised by malignant tumours, liver disorders, skin complaints and nervous-system ailments.

Metals not ordinarily toxic to humans can be highly dangerous to other living things. Foremost among these are tin and copper, both of which have been used as anti-foulants on boat hulls. One tin-based anti-foulant, known as tributyltin, which came into widespread use in the 1960s, has been described as 'the most toxic substance ever introduced into natural water'.[167] Its most severe effect is sterility in molluscs through the imposition of male sexual characteristics, a phenomenon known as imposex, that results from the disruption of the molluscan endocrine system. As little as half a nanogram per litre of water will cause female dogwhelks to start growing penises.

Imagine, if you can, the life of a deep-sea cucumber. Your role in the Gaian system is to dredge through the sediments of the ocean floor, many kilometres below the surface. Your world is eternally dark and frigid, and the rain of sediment from above is so meagre that it takes a thousand years to accumulate a depth of one millimetre on the sea floor. But then ships start ploughing the sea—ships whose hulls are coated with tributyltin, which is designed to flake away. The rain of flakes under the shipping channels soon resembles a slow-motion snow flurry, and the ocean floor is forever transformed. Although a total ban on tributyltin on ships' hulls came into effect in 2008, its effects will be with us for a long time, for the flakes are now present in marine sediments worldwide, and it's likely to influence the organisms that live in the sediments for a human eternity.

Like the metals that can poison us, some of the dangers of radiation come from a redistribution of elements we humans create between the three organs of Gaia. Today the Sun is our nearest naturally occurring nuclear reactor, and it supplies most of our energy (including our fossil fuels, the product of ancient captured sunlight). But around 1.8 billion years ago the concentration of radioactive elements on Earth was sufficient for the planet to develop natural nuclear reactors. The best evidence we have for these nuclear reactors comes from the Oklo and Bangombé regions of Gabon, in western central Africa, where sixteen separate natural nuclear reactors have been unearthed. They differed from the modern reactors used to generate electricity in that they burned more slowly—yet they continued to operate for millions of years.

These ancient nuclear reactors were discovered when France's Commissariat à l'Énergie Atomique began mining in the area. The uranium they recovered consisted mostly of Uranium-238: spent uranium. They found little Uranium-235, the type needed for nuclear reactors. At first they suspected that terrorists had somehow stolen the ore and replaced it with spent uranium from modern nuclear reactors. But studies subsequently showed that the Uranium-235 had been used up 1.8 billion years ago.[168] It had accumulated in algal mats in the estuary of an ancient river that flowed over uranium-bearing rocks. The algae absorbed the uranium, just as plankton absorb and concentrate radioactive elements today. Eventually the concentration was sufficient to start a nuclear reaction, which used up the Uranium-235 and killed the algae.

Why, then, are there no natural nuclear reactors on Earth today? Radioactive elements decay at a specific rate. The half-life of Uranium-238 (the time it takes for half of a quantity to decay into Thorium-234) is 4.5 billion years. So there is only half as much Uranium-238 on Earth today as when our planet formed 4.5 billion years ago. The half-life of Uranium-235, however, is only 713 million years. When the Earth formed, Uranium-235 made up around 33 per cent of all fissile uranium, but today it forms just 0.7 per cent. Thus our Earth is depleted in Uranium-235 to the point that natural processes cannot concentrate enough of it to trigger a nuclear reaction. Humans, however, by digging into the earth, can retrieve and concentrate Uranium-235, and revive a process long vanished from our planet.

While the dangers of this new nuclear age, with its problematic waste and potential for accidents, will long be with us, there is another more worrying legacy that is less often considered: the estimated 1740 tonnes of Plutonium-239 that was manufactured for use in nuclear weapons.[169] Plutonium-239 has almost no use except in nuclear weapons—its main purpose is to destroy people. Because it has a half-life of only twenty-four thousand years, before 1945 there had been no Plutonium-239 on Earth since the days of western Africa's natural nuclear reactors nearly two billion years earlier. Its sudden reappearance as a result of the atomic bomb program is as astonishing and troubling as would be the resurrection of a long-extinct dinosaur. And, like some resurrected movie monsters, it continues to grow and multiply in the darkness created by national secrecy. We don't know with any accuracy how much of the stuff exists, or precisely who has it. And if there's no way to account for it, we have little hope of safely disposing of it.

There are, however, ways to deal with the problem. A small percentage of the Plutonium-239 produced in nuclear reactors is burned as fuel, but currently that's an expensive process. A recent development, whereby Plutonium-239 is mixed with uranium oxide to form a mixed oxide fuel, may offer a cheaper alternative, but at present it seems that the only realistic option is to store it in secure vaults. And here the problem becomes one of politics as much as economics and technology. We need a global treaty to enforce the locking away of all Plutonium-239, which is a whole lot safer than having it scattered through countries such as Kyrgyzstan, Russia,

Israel and possibly North Korea. To cage the plutonium dinosaur safely would cost billions, but the money is less of an obstacle than political will.

In April 2010, the world moved closer to that goal. President Barack Obama convened the Nuclear Security Summit, which was attended by over forty world leaders. They agreed that the nations possessing nuclear materials must secure them within four years. It's clearly only a start—if an ambitious one—towards the eradication of Plutonium-239 and thus to the end of the threat of nuclear war.

As this treaty suggests, we may now be moving into an age where regulation of dangerous elements is inevitable, and not just for Plutonium-239, but uranium, lead, cadmium, mercury and carbon, to name just a few. Our best hope for dealing with many of these elements is the enforcement of cradle-to-grave responsibility, a move we've seen in other industries such as the auto sector. Mining interests will fiercely resist such regulation, for miners were born on the frontier and the laissez-faire spirit of the gold rush is alive and well among them. Civilising such powerful interests, which are deeply rooted in libertarian culture, is an extraordinary challenge—and already one element looks set to be the test bed of whether humanity can overcome such suicidal self-interests. That element of course is carbon.

Over the last two hundred years, humans have increased the CO_2 concentration in the atmosphere by a whopping 30 per cent.[170] This increase makes earlier human impacts on our atmosphere look trivial. The excess carbon has come from two sources: around 40 per cent from the destruction of Earth's forests and soils, and the rest from digging up and burning the dead—fossilised carbon in the form of coal, oil and gas. These fossil fuels began as plants, whose remains were entombed in Earth's crust as swamps and sediments from the ocean floor were buried deep underground. The carbon they contained was mineralised, and would have remained safely sequestered but for mining and drilling and burning. The great flux of carbon out of Earth's crust and into the atmosphere that has resulted from these activities has created the most urgent environmental challenge facing us. As we've known since 1859 when John Tyndall invented a machine capable of measuring the amount of radiant heat absorbed by various gases in the atmosphere, and so demonstrated the heat-trapping capacity of CO_2, humanity is tampering with Earth's sensitive thermostat. Almost every national scientific academy on Earth—from the Chinese to the Russian to the American, the Indian, Canadian and Britain's Royal Society—supports this view.

The carbon concentration in the atmosphere continues to grow at an ever faster rate. Two hundred years ago the atmospheric concentration of carbon (as CO_2) was around 2.8 parts per ten thousand. Today it's 3.9 parts—a level not seen for at least three million years—and even if we ceased burning fossil fuels today, it would take several centuries for life, the oceans and Earth's crust to re-absorb the excess. But that will not happen. Instead we're on track, if we do nothing, to increase the concentration to at least seven parts per ten thousand by the end of the century.

We've known, ever since the first report of the Intergovernmental Panel on Climate Change (IPCC) in 1988, that this rate of increase represents a grave danger, and with each new report more ramifications have been discovered. The most recent, issued in mid-2009, was based on the findings of a conference attended by two and a half thousand scientists in March of that year, and it makes alarming reading indeed. It sought to provide an update of the IPCC Fourth Assessment Report, which was released in 2007 (but was based on scientific observations only up to 2005).

With almost four years of new data, the conference was able to re-examine some of the IPCC's findings, and to test its projections. Among its key findings was that there was 'overwhelming' evidence that the burning of fossil fuels now threatened the development and wellbeing of human societies—a much stronger statement than anything provided by the IPCC. The scientists also concluded that the IPCC's Fourth Assessment Report median projections under-estimated the rate of climate change.[171] The oceans are a kind of global thermometer. Just like the mercury in a glass thermometer, the water in the oceans expands as heat trapped by the atmosphere is absorbed by it. Additional water is added from melting glaciers and icecaps as a result of the warming atmosphere. Furthermore, because the oceans are very large, they have a lot of inertia—they are slow to respond. The atmosphere, in contrast, responds quickly to temperature-changing factors, so there's lots of year-to-year variation. The average temperature of the ocean's surface is a more steady guide to the warming trend. It's concerning to discover that sea-level rise is tracking the upper limit of the IPCC's 2007 projections of more than a metre within ninety years. This would be a catastrophe for much of East and South Asia, the east coast of the US and parts of Europe, which are low-lying. And things could be far worse than that if a large ice shelf collapses and melts, for the sea could then rise even further.

The IPCC was more accurate in projecting Earth's average surface temperature, which is tracking within the scope of its 2007 projections. But it underestimated the rate of CO_2 accumulation in the atmosphere—real emissions for 2005, 2006 and 2007 were in excess of their worst-case scenario. The incidence of extreme weather events, the decay of the Arctic and Greenland icecaps, and the acidification of the oceans (caused by the absorption of CO_2 in sea water) are all proceeding at unanticipated rates. Indeed, the situation is now so severe that the report notes that 'global average surface temperature is unlikely to drop in the first thousand years after greenhouse gas emissions are cut to zero'.[172]

Today, the most important cause of these changes by far is mining and the burning of fossil fuels, which continue to increase and now account for around 80 per cent of all emissions. For the first two hundred years following the industrial revolution, emissions increased on an average of 2 per cent per year, but since 2000 they have increased by an average of 3.4 per cent per year, a rate that is driving atmospheric CO_2 levels beyond the worst-case scenario of the IPCC projections.

One of the most worrying aspects of the new report is confirmation that we had ample warning but did not respond. From the decline in the shell thickness of microscopic plankton in the Arctic Ocean due to acidification of the seas, through to a rise in sea level and increasing land and air temperatures observed to 2009, scientists predicted what was going to happen, even if they underestimated the speed of change. And the scientists' predictions for the future are grim. If we continue as we are for a few more decades, experts such as James Hansen believe that we're likely to trigger a shift to an ice-free Earth, which will eventually raise sea levels by tens of metres. In his book *Storms of My Grandchildren*, Hansen explains just how close we are to triggering a dramatic rise in the sea level, and what that implies. He argues that the heat imbalance of Earth—currently half a watt per square metre—is almost sufficient to destabilise the ice sheets of west Antarctica and Greenland.[173] The initial rise caused by a partial collapse may be small, but we'll have no idea how long levels will continue to rise or where it will end—whether with a one-, four- or fourteen-metre rise.

The effect on biodiversity is equally bad. If we continue as we are, within this century we stand at risk of exterminating up to six out of every ten living species. While we as a species won't become extinct, individual humans are likely to suffer greatly. James Lovelock believes that nine out of every ten of us living *this century* will die from climate impacts, leaving a population of just a few hundred million clinging to refuges in places such as Greenland and New Zealand. And of course that would destroy our global civilisation.

Towards the end of this book we'll examine how well humanity is addressing the problem of climate change. Now, however, we must return to contemplation of the human superorganism as it exists today.

5

OUR PRESENT STATE

CHAPTER 16

The Stars of Heaven

Remember Abraham, Isaac, and Israel,
thy servants, to whom thou swarest by
thine own self, and saidst unto them, I will
multiply your seed as the stars of heaven...
EXODUS 32:13, KING JAMES BIBLE

According to Thomas Malthus overpopulation was inevitable, because population growth is potentially exponential, while the means of feeding people increases only arithmetically. The outcome, he wrote in 1798, would be that:

> Epidemics, pestilence, and plague, advance in terrific array, and sweep off their thousands and ten thousands. Should success be still incomplete, gigantic inevitable famine stalks in the rear, and with one mighty blow, levels the population with the food of the world.[174]

There is no doubt that Malthusian logic rules the animal world, for unless populations are held in check by disease, predators or starvation, they continue to grow until they abruptly collapse. Consider the rabbits that ate Australia. From a few dozen European imports in the mid-nineteenth century, their numbers grew into the billions. Having eaten the heart out of the inland by the early twentieth century, they migrated desperately in search of food. The Australian government built a 'rabbit fence' to keep the hordes from the agricultural lands, but the corpses soon piled so high against it that they formed a ramp, allowing starving multitudes to pour onto the virgin pastures beyond. But even this reprieve could not delay the inevitable collapse.

Superorganisms such as ants seem, at first glance, to have avoided a Malthusian fate by making reproduction the privilege of the few. In the most highly evolved ant societies, such as the leafcutters', breeding is done by a single queen. Yet this does not allow the ants to escape the Malthusian trap. Were ant colonies not checked by diseases, predators and lack of resources, they too would soon cover the entire Earth, and then crash as did Australia's rabbits. So it is that superorganisms are constrained by exactly the same evolutionary forces that constrain us as individuals. It's this seemingly inexorable propensity of life to multiply and destroy that lies at the heart of the Medea hypothesis, which states that life itself periodically brings about the destruction of life, and that long-term ecological stability is impossible.

There is a Medean aspect to the way reproduction in our own species often pits the interest of business, religion and the state against the individual. In developed countries most governments discourage stable or shrinking populations because they threaten the tax base and national prestige. Governments deploy an array of incentives to influence our reproduction, ranging from limiting access to pregnancy termination to banning treatments such as the early-stage abortion drug RU486, to cash payments for births and a redistribution of income to large families. Business lobbies are also disdainful of low rates of population growth because they threaten profits, and religion too remains an important influence as the more adherents a faith has, the greater the prestige of its pope or mullas. The population policy of the Catholic church is well known, but a very different example concerns Saudi Arabia. One of the richest countries on Earth, it demonstrates that affluence doesn't always induce smaller family size. Saudis have a fertility rate of 5.5 children per woman (compared with 2.7 for Earth as a whole, and 3.5 for the Middle East and North Africa), which presumably reflects the limited education and choice the kingdom affords its women, and the fact that men have disproportionate power and can force women—at no cost to themselves—to bear the consequences of their desire for greater fertility. Such anomalies, however, are increasingly rare, and there's little chance that they will affect humanity's demographic future overall.

If all else fails to move a nation to increased reproduction, there's always immigration. A few highly affluent nations, including the US, Australia and Canada, have enhanced their growth through the recruitment of new citizens from abroad long after their populations have moved to low-growth rates of fertility. These countries share a common history. All are frontier cultures where until recently national destiny seems to have consisted of a quest to populate an eternal frontier—a situation in which, as we've seen, our Medean face shines most brightly. But that phase of their development is now over, and it is critically important for the future of Earth that these nations understand this.

Immigration brings many benefits—not least the establishment of a shared global identity—and if immigration policies in places like the US, Canada and Australia are tied to strong efforts at reducing environmental impacts, there will be no problem. Alas, these nations are precisely the ones with the worst records in terms of greenhouse-gas emissions and consumption of resources. For those who advocate increased immigration, sustainability is not optional; it's imperative. Yet it's often the very individuals who oppose environmental initiatives who also welcome population growth. They want to have their cake and eat it too.

The man who first saw the dangers of overpopulation, Thomas Malthus, lived when innovations in western Europe were overcoming causes of premature death (particularly in childhood), yet women were having as many babies as ever. Had things continued that way a Malthusian catastrophe would indeed have been inevitable. As it turned out, however, medical science and society continued to change and evolve, allowing us to limit reproduction with ease. But has the change been swift enough to deliver us from Malthus' nightmare? Every two years the population division of the United Nations Department of Economic and Social Affairs publishes a projection of global population, the latest of which was completed in 2008 and published 11 March 2009. In each of the previous UN publications, Earth's population was projected to be higher in 2050 than in the previous estimate. In 2008, however, something remarkable happened: the population estimate was lower than that of the previous (2006) projection by around forty million.[175] How had this come about? It's an important question, for if starvation and disease were the cause, then we could see evidence of Malthusian logic. But if not, something else must be at work.

The details of the report are intriguing. Because of a slight rise in the fertility rate of some wealthy countries, the total human fertility in the 2008 projection is actually higher than that of 2006—up to 2.56 children per woman from 2.55. And the report assumes that humanity will be substantially successful in its battle with AIDS, predicting thirty million fewer people dying from that disease than was previously estimated. It's logical to think that these factors would result in a higher 2008 population projection for 2050 than the projection of 2006. But another trend counteracts them—a sharp decline in fertility in the world's forty-nine poorest countries, brought about by increased affluence and education and by access to family planning.

As with any projections, the UN report deals with probabilities, not certainties. The most likely outcome is a total global fertility rate (the average number of children born per woman) in 2050 of just over 2 children (down from the present 2.56), which would see the human population peak at 9.15 billion by mid-century. There is a smaller chance that by 2050 women will have (on average) only 1.5 children, in which case our population would reach just eight billion. There's also the possibility that the birthrate will remain unchanged from today at 2.5, in which case there'll be 10.5 billion of us in 2050 and our population will continue to increase for decades longer.

There is ample evidence in the poorest countries of a desire for family planning, as well as a colossal lack of ability to achieve it. Our common future rests upon realising such aspirations, and if we succeed, within a span of just 150 years all of humanity will have passed through a demographic transition never before seen in human history and without parallel in any other species.

So what is the demographic transition? In its most basic form it explains how both death rates and birthrates decline as a country becomes industrialised and its population becomes more affluent and educated. The idea was developed in 1929 by American demographer Warren Thompson, who based his theory on two centuries of population statistics from the most developed countries. Modern demographers divide the demographic transition into four stages. At first both birth and death rates are high, for people are vulnerable to innumerable threats and have little control over their fertility. Then, as a country develops, the death rate drops because disease is brought under control and health improves. This is the stage Malthus lived in, where population growth is rampant. In stage three, however, birthrates drop too—because of access to contraception, an increase in the cost of education and a decrease in the value of children as workers (making it more expensive to have large families). In the fourth stage, which is characteristic of highly developed countries, the birthrate drops below replacement and in the absence of immigration the population eventually shrinks.

There's no guarantee that the demographic transition in the developing world will resemble that of the west. China's demographic transition, for example, was very different. Following the encouragement of large families under Mao the nation faced a disaster, and a one-child policy was implemented over much of the country. This brought about an extremely swift demographic transition—China now has an average birthrate of 1.6, but only at the cost of highly restrictive policies including enforced late-term abortions. Because of a desire for sons, China's population policy has also created a significant gender bias in recent generations. An analysis of births for 1985–86, for example, found that 10 per cent more males were born than females, which indicates that, for that year alone, half a million female infants were missing from the population.[176] These factors indicate that in coming decades China will experience a dramatic aging of its population.

Demographers debate whether there will be a fifth stage in the transition, to an even lower reproductive rate than is seen in developed countries today. In the Russian federation the population is shrinking significantly—having gone from 146,670,000 in 2000 to 140,367,000 in 2010—and is projected to be just 128,864,000 by 2030.[177] Yet some countries with very low birthrates, such as Italy, will continue to grow for several decades because of immigration. By comparing Italy with Japan (with fertility rates of 1.31 and 1.34 respectively), you can see what a difference immigration makes:

Italy (which has immigration) will continue to grow for a decade or two longer, while Japan (which has almost none) is projected to drop by ten million by 2030.

From an evolutionary perspective, the demographic transition is a profound enigma. Clearly most people could, even if it cost them some hardship, have larger families, and yet they choose not to. Evolution by natural selection should see us optimising our reproductive potential just as most humans and other animals have done throughout history. If our genes were still in full control of our bodies, this is what natural selection would surely compel us to do. But a new force has come into play—the mneme (or belief)—and it is more powerful than anything that has come before. One of the most important mnemes of the twentieth and twenty-first centuries is that individuals, both male and female, are important and have a right to improve their own lives. In this instance our personal interests and those of the environment coincide, paradoxically benefitting Gaia.

With cheap and convenient contraception widely available, the mnemes of a large portion of humanity have immense power, and it seems that they've engineered a profound reversal of evolutionary principle as it's understood in brutal Neo-Darwinian terms. At its heart the demographic transition represents the triumph of the individual against the tyranny of the selfish gene. A stabilisation followed by a decline in human numbers makes a sustainable future possible, and if we achieve it then Wallace's vision of a perfection of the human spirit may eventually be realised.

Many argue, however, that the planet is already overpopulated. It is reasonable to question whether Earth can sustainably provide resources for nine billion people—the 6.8 billion of us alive today use 30 per cent more resources than Earth can sustainably provide, and that's with many living in poverty.[178] I believe that Earth can support nine billion, at least for a few decades, and shortly I'll tell you why. But what if I'm wrong? What should we do about it? We could, and should, hasten the demographic transition as much as possible, consistent with human dignity and rights. But can more than that be done? Who would you ask to get off the planet if the need arose? The truth is that, if we wish to act morally, then we can reduce our population only slowly. So while it's important to focus on population as a critical element in the long-term solution of our problems, we cannot make it our only focus as we seek to deal with immediate challenges such as our destabilising climate.

CHAPTER 17

Discounting the Future

At every stage there is the threat of
mutiny, of rebellious individualism that
might destroy the collective spirit.
MATT RIDLEY 1996

We now come to what many believe is, after population, the greatest obstacle in our path to sustainability. It's not one of the seven deadly sins, though like them it was instilled in us through evolution by natural selection. But it is so little recognised that we don't even have a word to describe it. Among sociologists it's known as discounting the future—taking short-term gain even though doing so might cost us immensely in the longer term. At its most extreme, it manifests with mnemes expressed in ideas like 'I've nothing to lose' or 'I've no future'. While these sentiments may lead to a sort of moral paralysis, our genes do not give up. Instead, in their search for immortality, they seek to squeeze as much as possible out of us before we crash and burn. And that can be very bad for us, our society and our planet.

We are all familiar with the pictures in the news: youths in balaclavas captured on CCTV robbing service stations for a few dollars; young men stabbed or shot dead over some seemingly minor affront. Why is it that some of us do risky things that could cost us our futures? Psychologists Margo Wilson and Martin Daly of Canada's McMaster University have spent an academic lifetime studying the problem, and think the answer lies in the rates at which individuals discount their future. We do not have to undertake rigorous studies to understand aspects of this problem. The mother of one of my friends is in her nineties and has a keen interest in the stock market. When her son suggested that she purchase a particular stock with medium-term prospects for profit, she looked at him disdainfully before saying that at her age she didn't even buy green bananas. But our responses are often far more complex than that. In recent years psychologists have devised ingenious tests to determine how our 'discount rate' varies, and some of their results are truly surprising.

The most common tests involve asking volunteers whether they would be willing to wait for a reward. An interviewer might say, for example, that the volunteer either can have $100 now or can nominate a sum they'd be willing to accept if they had to wait a year for payment. Waiting of course involves risk: the researcher may go broke, for example, or the volunteer might die before they get paid. It turns out that the amount required to induce people to wait varies with age and sex. Men on average require a larger amount than women. But, overall, the payment for patience is high, nearly $500—five times the immediate payment.[179] By comparison, investing the immediate payment of $100 with a bank for a year would yield $10 or less in interest, indicating that our natural discount rate is steep indeed.

The conclusion most of us would draw from such studies is that, while humanity might be impatient, men are more impatient than women. Social scientists, however, look at things in evolutionary rather than moral terms. It's not just humans who have a naturally high future discount rate. Ingeniously devised tests of birds reveal that they also have to be offered large inducements to wait for rewards. Clearly, both our brains and those of birds were forged in an environment where survival was uncertain, and if you are unlikely to see tomorrow, why not take what you can get today, even if it means foregoing a far greater reward in the future?

The tests described above are examples of game theory, which in recent decades has become enormously important in the social and political sciences, as well as in economics and biology. Indeed some researchers now refer to it as the universal language of the social sciences.[180] Typical of the problems investigated is that outlined in 1651 by Thomas Hobbes in his book *Leviathan*. Writing during the time of the English Civil War, Hobbes was interested in demonstrating what happens when free individuals interact. Many, he thought, might cooperate. But there will always be some who seek a free ride. So, for example, a person might get others to help build his house, but then refuse to assist them in return. This would make the defaulter worry about revenge. He might anticipate that someone will burn down his house. So fear prompts the defaulter to pre-empt this by killing the person he defaulted on. Thus is chaos born. The only alternative, Hobbes argued, is the rise of a dictator who will punish bad behaviour.[181] By allowing us to investigate the nature of cooperation under various circumstances, game theory sheds light on the validity of Hobbes' hypothesis.

In the social sciences, game theory experiments are often of the laboratory type, and typically take as their subjects students wanting to earn a few dollars. The holy grail of some game theorists has been the demonstration of altruism. So far the jury is out on whether humans can be relied upon to be truly altruistic towards non-family members, but game theory has taught us as much as any laboratory experiment can about the circumstances in which individuals do cooperate and treat each other fairly. One such experiment involved giving subjects sums of money and instructing each of them to split the money with a person unknown to them. If the stranger rejects the sum offered, both go home empty-handed; if not both keep the shared amount. Generally, the giver offers around a third of the total, and this is usually accepted. But if it drops to around a sixth, the offering is frequently rejected, leaving both with nothing.

Such games are important to those seeking to shape better societies. Among the many aspects of human behaviour elucidated by game theory is the notion that we hate losing money more than we enjoy gaining it—an important insight for those promoting efficient use of resources, for example.[182] As we shall see, game theory has also been applied to nations seeking to broker a climate treaty, revealing interesting behaviour. But what are the factors that influence us to discount our future?

Generally, the less security we have, the more heavily we discount our futures. Heroin addicts (who live with the daily risk of death by overdose or disease) have future discount rates twice as high as those of comparable non-addicts, which may explain their willingness to participate in high-risk activities such as prostitution and crime.[183] But the starkest example of circumstance influencing the discount rate concerns homicide rates among young men. As the Canadian husband and wife research team of Martin Daly and Margo Wilson put it:

> Homicide rates are at their highest among men with the
> least to lose...In our own research on homicide in Chicago
> neighbourhoods, income inequality was, as usual, an excel-
> lent predictor of homicide rates, but we found an even better

predictor: local life expectancy (with the mortality effects of homicide removed to prevent circularity)…What this suggested to us is that dangerous competitive behaviour that entails an implicit disdain for the future is exacerbated by cues that one lives in the sort of social milieu in which one's future may be cut short…Rather than vilifying those who discount the future as myopic or lacking in self-control, we think it is both more accurate and more fruitful to hypothesise that steep discounting characterises those with short life expectancies, those whose likely sources of mortality are independent of their actions, and those for whom the expected fitness returns of present striving are positively accelerated rather than exhibiting diminishing marginal returns.[184]

What Daly and Wilson are saying is that disadvantaged young men kill other young men over small sums of money and minor insults because they have a high chance of dying in the near future anyway, and with nothing but their social status and a few ill-gotten dollars to spend to impress a girl, doing that right now becomes the most precious thing in the world. In evolutionary terms, it's worth risking death in a knife fight, or a lifetime in jail, to get a woman pregnant. Evolution is, as Richard Dawkins describes it, a blind watchmaker that cares not a jot for us as individuals—pregnancy being merely a passport to potential immortality for selfish genes.

Young women can also sharply discount the future. Teenage girls with life-threatening illnesses have higher than average rates of pregnancies and incidences of STDs, indicating that despite their ill health they are partaking in risky sexual activity. Again, the body is maximising the chance of immortality for its selfish genes. Indeed one of the best documented effects of living with the constant threat of death in wartime or pestilence is the increase in casual sexual activity. Samuel Pepys' diaries, documenting his adventures in seventeenth-century London, were written during a period when the great fire, the plague and the Dutch all threatened the city. And the worse the threats got the more sex Pepys enjoyed, for both his own libido and those of countless women seem to have been piqued by the peril.[185] In more recent times this tendency was a cause for concern during World War II for British authorities, who fretted over the spread of venereal disease during the blitz, and provided condoms and education to combat the threat.[186] It seems that, as our prospects diminish, the power of our selfish genes over our mnemes increases.

The tendency to discount the future helps explain why people sometimes act to destroy their environment, whether by cutting down rainforests, continuing to pollute the atmosphere or destroying biodiversity. And people without prospects are created in a number of ways—through grinding poverty, through greatly unequal societies and through war, famine or other misfortunes. If you're concerned about our future, it's not just desirable that we eradicate poverty in the developing world, create more equal societies and never let ourselves fight another war; it's imperative, for the discount factor tells us that failure to do so may cost us the Earth.

CHAPTER 18

Greed and the Market

Business is 'just business': a scramble for
profit. Right? Well that might describe
crime; it certainly doesn't describe business.
Ethics aren't just important in business.
They are the whole point of business.
RICHARD BRANSON 2008

Neoclassical economists believe that economic systems are driven by a trinity of human rationality, greed and equilibrium. While people may not always act rationally in economic decision-making, or through mere greed, there is no doubt that greed and its neighbour selfishness are near-universal human characteristics.[187] So does this mean that humans and their market systems must inevitably, Medea-like, imperil the planet? Brains are notoriously selfish organs. They give themselves priority access to everything they require—from blood-flow to warmth, nutrients and oxygen. During times of bodily stress, our brains will shut down one organ after another, even to the point of damaging them—before depriving themselves. Brains are also greedy. They make up just 2 per cent of our body weight, yet take 20 per cent of the energy we use.

Greed and selfishness can be thought of as essential characteristics of command-and-control systems: the Fat Controller is not merely a fiction from a children's story; he's fat precisely because he controls. Yet neither fat controllers nor brains destroy the systems that they are part of. As our influence on Earth systems grows, accommodations must be made that permit us greedy, selfish beings to live in balance with the rest of nature, and that includes in our market system.

Markets are essential to society's prosperity and dynamism, but unconstrained they can become juggernauts capable of crushing humanity. In eighteenth-century Europe, wives and children were bought and sold at the auction block, and paupers could sell their teeth to be set in the heads of rich men. Society's standards may have since changed, but they still permit some to behave in ways that harm our common futures.

The justifications of CEOs seeking to profit at the cost of others are, in my experience, remarkably uniform, running along the lines of 'everybody does it, so why shouldn't I?' This is often followed by something like 'well, try and stop me then'. And true it is that some of their most damaging activities, including unrestricted emissions of greenhouse gases, remain legal in many jurisdictions and are hard to stop. Furthermore, their activities are given a veneer of respectability by philosophies such as those of the neoclassical Chicago School of economics, which sees a minimal role for government regulation in the marketplace. There is more than a passing similarity, incidentally, between neoclassical economics and Dawkins' selfish gene theory. Both describe idealised frameworks which can be powerfully explicatory. But when they become universally dogmatic, ideologies have the power to erode our capacity to value one another, and so threaten to destroy the common endeavour that is our global superorganism.

Researcher Robert Frank and his colleagues have discovered

a remarkable thing about people trained in neoclassical economics. In a number of tests, including a game theory test called the prisoner's dilemma (which measures a person's propensity to trust in and cooperate with others), they found that neoclassical economists are more likely than other people to betray their partners.[188] Frank also discovered evidence that 'differences in cooperativeness are caused in part by training in economics'.[189] Is it the case that those who subscribe to theories of absolute selfishness as an evolutionary imperative themselves become more selfish? As science writer Matt Ridley comments:

> The virtues of tolerance, compassion and justice are not policies towards which we strive, knowing the difficulties upon the way, but commitments we make and expect others to make—gods we pursue. Those who raise difficulties, such as economists saying that self-interest is our principal motivation, are to be distrusted for their motives in not worshipping the gods of virtue. That they do so suggests that they may not themselves be believers. They show, as it were, an unhealthy interest in the subject of self-interest.[190]

As we've seen, selfishness is not always negative. A certain kind of selfishness propels the demographic transition. And were we entirely unselfish, our societies would be little different from those of the ants. What Ridley is talking about here is the kind of selfishness that erodes our common bonds and futures, and that erodes the value we put on the fact that our interconnections are complex.

How is it that neoclassical economists may become selfish in ways that harden them to the needs of their society? In addition to the influence of training identified by Frank, it seems possible to me that their moral sense can become corrupted through their work—by mnemes passed on from their professors and colleagues.

Like the CEOs whose business activities damage others, the economists, through self-selection of friends, may come to believe that everybody (or at least everybody who counts) thinks just like they do. And so an environment is created where selfish individuals are selected for. This then influences government and society at large because our civilisation requires economic planning and that planning must be informed by economic theory. Writing before the advent of neoclassical economics, Adam Smith had this to say of business interests:

> The proposal of any new law or regulation of commerce which comes from [the business community] ought always to be listened to with great precaution, and ought never to be adopted till after having been long and carefully examined, not only with the most scrupulous, but with the most suspicious attention. It comes from an order of men, whose interest is never exactly the same with that of the public, who have generally an interest to deceive and even to oppress the public, and who accordingly have, upon many occasions, both deceived and oppressed it.[191]

Economists as well as social scientists employ a discount factor. In business it's used to compare costs and benefits of investment decisions in ways that determine the timing of such decisions: for example, a company needing to replace a piece of machinery within a few years will ask whether it's more cost-effective to make the purchase now, or in two years' time. In order to assess this it will look at many factors, including the inflation rate and whether the piece of machinery is likely to become cheaper or more expensive. If the discount factor so determined is high, the investment will be deferred.

This is a perfectly reasonable and acceptable exercise. But

when a discount factor is applied to problems with environmental or personal consequences, things are not so clear. The economic discount factor (which represents the amount by which cash flows are spread over time) is, for normal business purposes, almost invariably set high. As Professor Darren Lee of the University of Queensland puts it, the consequences of applying a high discount factor is that 'long-term environmental threats are not addressed by markets in part because, from a financial perspective "anything" that happens that far out simply has no tangible value in today's dollar terms whatsoever'.[192] Economists who argue that we should apply a high discount rate to the analysis of spending on climate change are essentially arguing that it's best to leave future generations to fend for themselves, because expenditure in future will combat threats far more cost-effectively than anything we can do today.

Lord Nicholas Stern used a near-zero discount factor (tantamount to using no discount factor at all) in his cost-benefit analysis of addressing climate change. Partly as a result, he argued that large amounts of money need to be spent now to achieve effective action.[193] He was widely criticised for this. But was he right? One way of understanding the terrible consequences of applying a high discount factor to such problems is to imagine applying it to yourself in a medical matter. Imagine that you have cancer, and the doctor tells you that you have five years to live, but that a novel and expensive treatment is likely to save your life. An economist applying a high discount factor, and looking at the treatment as an investment decision, might decide to wait before shelling out, on the basis that the cost of the medicine may decrease greatly, so he'd only be wasting money by buying it now.

Climate projections and medical diagnoses are both probabilistic in nature. When the doctor says that a patient has five years to live, he means that, on average, people with a similar condition live for five years following diagnosis. Some, however, die much sooner, while others may live on for decades. When climate scientists say that, unless something is done, a catastrophic climate impact is likely by 2050, they mean that this is the most probable outcome—but there's a smaller chance that the impact may be felt in the next few years, or perhaps not until after 2050.

One specific objection to Stern's near-zero discount rate was that while the climate impacts may not be felt for decades, the money must be spent now, so it involves a transfer of wealth from this generation to a future one. In the case of our current climate projections, however, we need to keep in mind the possibility of a catastrophic outcome in the near future—one that will affect this generation. We also need to acknowledge the fact that, once climate change has progressed beyond a tipping point, no amount of investment can halt it. Such possibilities urge immediate action, as a kind of insurance payment, and would seem to justify Stern's low discount rate.

Health and environment are not the only circumstances where normal (high) discount factors can be dangerous. If the steep discount factor that prevails in most business decisions guided our long-term expenditure in education, for example, we might decide to invest less in our children—say by lopping a few years off school—on the basis that the bulk of students would be minimally disadvantaged, and that in any case the investment has a long payback period and an uncertain return. Even in building and agriculture a high discount factor can lead to poor outcomes. We would have no Pantheon, for example, or any ancient buildings at all, had our typical discount rates been applied by previous civilisations; nor would we have long-term soil fertility, because we would not invest sufficiently in maintaining it. By applying a high discount factor in such circumstances we risk leaving behind an ignorant and rotted world, because thus applied it does to us collectively what discounting the future does to individuals—induces us to take short-term gains at the cost of inflicting enormous, and possibly terminal, long-term pain on our global superorganism.

The global economic crisis has brought the short-term nature of markets into sharp focus. Part of the result can be seen in the current focus on the remuneration of CEOs, especially those taking stupendous bonuses despite driving their companies to major losses. One remedy to this culture of greed is the paying of bonuses and salaries on the basis of company performance over a multi-year period, thus diluting the focus on short-term profits. Generation Investment Management (GIM), established by Al Gore and David Blood (with its nickname of Blood & Gore), invests capital in ways that benefit future generations. When founding GIM in November 2004, Gore said that:

> Transparency, innovation, eco-efficiency, investing in the community, nurturing and motivating employees, managing long-term risks, and embracing long-term opportunities are integral parts of a company's enduring capability to create value. Business leaders who align their business strategy and technical development with sustainability and social accountability will deliver superior long-term results to shareholders.[194]

And they walk the talk, for partners at GIM are paid according to how investments perform over a three-year period.

The issuing of restricted stock grants as part of a CEO's remuneration has also been used to ensure a longer period is used to calculate profitability. In other cases, additional restrictions have been applied, including forfeiture of the shares if stock prices fall below their initial value after three years.[195]

New thinking is also focused on the problem of externalities. An externality is an effect or a by-product that is not accounted for in the costs of that business activity. A good example is the emissions produced by a coal-fired power plant. As Steven Levitt and Stephen Dubner say in their book *SuperFreakonomics,* an externality is 'an economic version of taxation without representation'.[196] We experience an externality every time we breathe polluted air or suffer from the effects of climate change. A big problem with externalities is that any company benefitting from them (by not paying for them) is likely to be more profitable than a similar company not benefitting from them. This means that an unethical investor is likely to make a larger profit than an ethical one, at least in the short term. Where externalities have grave environmental consequences, such investments may cost humanity dearly. There are only two solutions to this problem: government regulation (which requires that companies not produce the externality or pay a price for producing it) or a different investment model. Both face huge obstacles, but new models of investment are making modest advances.

One market-based means of dealing with externalities evolved from the old 'war bond' philosophy of utilising savings to help combat a common enemy. In Sweden the Skandinaviska Enskilda Banken has joined with the World Bank to raise 'green bonds' for climate change mitigation and adaptation work. The triple-A-rated bonds have a six-year maturity, and the interest rate payable is 0.25 per cent above the Swedish government bond rate. As of mid-2008, US$350 million had been raised from Swedish institutional investors—a small but promising beginning.

A second initiative, which has been suggested in the US, is the creation of a Clean Energy Bank. To be modelled on the US

Export-Import Bank, it would provide long-term finance for the provision of clean energy infrastructure. And recently an even more ambitious plan has been raised—the creation of a global fund whose sole purpose is the financing of projects designed to protect our global commons.[197] Operating on a similar model to the US Federal Reserve, it would ensure sufficient capital liquidity so that we do not need to deviate from an emissions-reduction trajectory designed to avert dangerous climate change. Funding would come from governments, and the agency would operate as a multilateral body. In summarising the kind of capital market that would be capable of addressing climate change, James Cameron of Climate Change Capital (an investment management company) and David Blood of GIM list the following principles: a long-term perspective, good governance and transparency, and cooperation, which are precisely the same principles required of markets if we are to improve the quality of our stewardship of Earth.[198]

One special case of an externality concerns the destruction of biodiversity. Here it's not youths in balaclavas that are trashing our world, but calculating gentlemen with clipboards. Imagine a business that owns a forest. It may determine that a certain return can be made from the forest if it's managed sustainably, but that a greater financial return can be had from destroying the forest and investing the cash made from doing so in the market. In this case, the impact of the forest's destruction on our living Earth is not accounted for in the transaction—it is an externality. But a strange valuation related to the discount factor is also at work. Basically, the value of the forest increases too slowly, relative to returns on investment in the market, for it to be considered profitable. Thus the decision to destroy the forest is entirely rational in an economic sense, yet if everyone accepted such inducements there would be no market, as our society would be overwhelmed by the disruptions

caused to Earth's climate and other systems. One way of dealing
with such comprehensive market failure is to remove high-value
forests from the economic realm altogether, which is what those
arguing for a cessation of logging of old growth forests seek to
achieve. Another is to legislate so that owners of forests use them
only sustainably.

Investors can help civilise markets. Shareholder activist Robert
Monks developed the universal investor theory, which argues that
because the largest investment funds hold shares in a wide diversity of businesses, their optimum strategy is achieved if society and
its supporting environment are in good health.[199] In other words,
the interests of the largest funds (universal investors) are aligned
with the interests of society and the environment as a whole. In
2007 institutional investors (large investors that own many classes
of assets, and thus are likely to be universal investors) owned
76 per cent of the thousand largest US companies, and thus are in
an extremely powerful position to influence the future infrastructure and behaviour of corporations.[200]

On the occasion of his fiftieth birthday, Monks presented a gift
to Harvard, his alma mater. It consisted of a letter, which read in
part:

> Harvard has become an 'owner' of virtually all of those
> enterprises whose collective functioning impacts life on
> Earth perhaps more than any other institutions. The
> question is the extent of Harvard's responsibility as owner.
> What is Harvard doing now? Does she ensure optimum
> value? What should she do in the future?

In answering his own rhetorical question, Monks went on to
point out that:

> Harvard has developed a very worldly competency to increase the asset value of its investments. Can one seriously object to her taking worldly responsibility for some of the consequences of her investments? After all, Harvard is one of the largest owners of public corporations; she is not a stranger to their impact on society. Institutions cannot simply define problems as being external to their mission; at some place 'the buck stops' and some institutions have to begin to accept responsibility. The alternative is a chaos where the most difficult problems are simply ignored by those best qualified to help, left to fester in the certainty that they will become toxic. Harvard must face reality. Our great university exists not in a tower, but in the real world—where businesses have a costly impact.[201]

I'm occasionally asked to speak to the managers of super-annuation/pension funds about the future risks that climate change may bring. I begin by explaining that the principal cause of greenhouse-gas pollution is the burning of coal, and that, if nothing is done about it, within a few decades climatic instability is likely to begin to threaten our economic system. This is a time frame meaningful to fund managers because the majority of their existing members will be drawing on their investments by then. I then ask for a show of hands to indicate the funds that are investing in coalmines, coal-fired power plants or other emissions-intensive industries. The hands show that most funds have such investments despite the fact that excellent scientific evidence indicates that these are the very machines that will destroy the wealth of their younger contributors by the time they come to call on them.

Thanks to Monks and a group of proactive fund managers, an initiative known as Principles for Responsible Investment was established in 2006. One program, launched in 2009, seeks 'to explore how the most economically harmful environmental, social and governance externalities involving corporations can be identified and then reduced through collaborative shareholder engagement with the owners of these corporations'. Issues of interest include greenhouse-gas emissions, biodiversity and eco-system services, resource use and efficiency, water use, food security, corruption, education, and health and social issues. By March 2009 the program had 470 signatories with $18 trillion under manage-ment. The work is aided by an online activist group known as Proxy Democracy.[202] Supported by foundations that are themselves interested in becoming responsible investors, Proxy Democracy assists investors in making decisions by publicising the past and intended votes of institutional investors, thereby making trans-parent their real commitment to a more sustainable future.

The universal investor theory is intriguing from a biological perspective, for it suggests we may be about to take a significant further step towards superorganism integration. As we've seen, individual ants use the good of the colony, rather than their own individual benefit, as the yardstick by which to set their discount factor. Perhaps, as humanity addresses the market failures that have led us to discount our own futures so steeply, we too will increasingly consider the survival of our own superorganism—our global civilisation—as the yardstick by which we establish an appropriate discount rate and profit horizon within the market system. Indeed, what the universal investor theory seems to tell us is that the markets that are so central to our superorganism should be regulated so as to ensure the future of Gaia, even where this means restricting short-term benefit to corporations and individuals.

CHAPTER 19

Of War and Inequality

The conditions existing in the world
today force individual states, out of
fear of their own security, to commit
acts which inevitably produce war.
ALBERT EINSTEIN 1945

As societies have grown more complex over the millennia, so they have grown more internally peaceful. Yet conflicts between these internally peaceful entities have escalated, even as the number of political power blocs has declined, until by the 1960s war between two of them could threaten the entire Earth. Today those power blocs have largely dissolved, and we muddle our way forwards, a species more united than ever, yet still with little structure for coordinating nations. In this new world, can the social glue that binds us keep the peace?

In his landmark book *The World Is Flat* journalist Thomas Friedman argues that peace will prevail as an inevitable consequence of trade. As he puts it:

> People embedded in major global supply chains don't
> want to fight old-time wars anymore. They want to make

just-in-time deliveries of goods and services—and enjoy the
rising standards of living that come with that.[203]

Stephen Green—the chairman of HSBC (one of the world's
largest banks), who also happens to be an Anglican priest—doubts
Friedman's assertion, as global supply chains (perhaps not 'just-in-
time' but nonetheless well developed) prevented neither of the world
wars.[204] We must remember, he says, that war is ultimately a political tool upon which business has a limited effect. One thing I'm
certain of is that the cost of war to highly developed nations—and
particularly to the large metropolises that lie at their hearts—is
only increasing.

We often think of future wars in apocalyptic terms—nuclear
weapons slamming into city centres and such like. But our modern
cities are so brittle that far less spectacular attacks could bring them
to ruination. In this they are very different from the London that
withstood the blitz. Today's cities rely upon highly sophisticated and
easily disrupted technology to deliver water, food, fuel and power
to populations of ten million or more, in a just-in-time manner.
Imagine a city like New York or London without a functioning
electricity grid. People living in high-rise buildings would probably
be trapped. With water pumps not working, both the removal of
sewage and the supply of clean water would immediately become
problems. Communications would be cut, traffic flows and rail
services paralysed, and with no refrigeration food would quickly
spoil. At night the streets would be plunged into darkness. Local
generators might keep hospitals and other vital infrastructure
going for a while, but within weeks the city would have to be
abandoned. Where would the millions go? If the disruption were
sufficiently prolonged it's fair to question whether the city would
ever be reoccupied.

As Friedman's argument implies, the disruption to trade that is an inevitable consequence of war is almost as great a threat as war itself. We are all now dependent upon an efficient global trade in goods like oil, coal, food, metals, medical equipment and much else on a colossal scale. And it's not only our cities and trade that are vulnerable to war, but our living planet as a whole. Its wellbeing is dependent on treaties and agreements that would break down in the face of war. Imagine the impact on the access to water from rivers flowing between warring states. Imagine the unconstrained production of toxic chemicals and greenhouse gases, and the overall resource extraction that would follow a war in which nations set aside their mutually agreed rules. In earlier conflicts, protected species such as wild-living European wisent fell victim; today all of Gaia could suffer, leaving a crippled planet in the hands of warlords.

In a globalising world the nature of conflict must change, for ultimately there will be no 'other' to fight. Future conflicts may be more akin to civil war, or even organised crime. This, incidentally, does not mean that conflict will be less bloody, for weapons are becoming more potent and readily available. It's worth remembering that on September 11, 2001, a few dozen fanatics killed thousands, forced the cessation of all air travel in the US for three days and damaged the world's largest economy.

As nations collapse, opportunities for the boldest increase. The people of Somalia are some of the poorest on Earth, earning an average of just US$600 per year, yet illiterate young Somalis have been able to capture and hold hostage some of the largest vessels ever built. The first attacks occurred in the early 1990s and were restricted to coastal waters. By 2008 Somali pirates, armed with fast boats, assault weapons and rocket launchers, were ranging almost a thousand kilometres out to sea, and collecting around $80 million a year in ransoms. Efforts to counter them, which involve the navies of twenty-six countries and cost billions, remain largely ineffective.

The power of motivated yet impoverished youth to threaten the relatively tranquil developed world will only increase in future, for the distances separating us are shrinking, our knowledge of each other is growing, and more dangerous technology is becoming more widely available. When I began fieldwork in Papua New Guinea in the early 1980s, tribal war was a highly ritualised affair carried out with bows and arrows. Today assault rifles are used, and the enemy is not always the traditional one, but can also include tourists and local workers. Leaving Papua New Guinea in its slough of corruption, with its often impoverished and disenfranchised people robbed of hope, is no more an option for the world than leaving Somalia or Gaza to its fate. Ultimately, our only hope of influencing young men with high discount factors is to empower their societies and families, and so give them something to live for.

For these reasons, poverty has to be everyone's enemy in a globalising world. The last few decades have seen the most astonishing progress in lifting the entrenched poor out of their misery, and our future depends on hastening this trend. In 1985 the average mainland Chinese worker earned just $293 per year, but by 2006, little more than two decades later, average earnings had risen to $2025. India is on the brink of a similar lift, and already this decade the proportion of Indians living in poverty has declined from 60 per cent to 42 per cent.[205] Some may doubt that such a transformation is possible in sub-Saharan Africa, a region afflicted by dictatorships, insupportable population increase, disease and poverty in almost equal measure. But Africa too is changing: the number of democracies in sub-Saharan Africa increased from four to eleven between 1995 and 2005.[206] Immense problems remain—Kenya and Zimbabwe are just two examples of countries slipping backwards.

But on the plus side the African Union is finally beginning to play a role in resolving conflict, per-capita incomes have been steadily growing (though too slowly), and abject poverty is declining in many areas.[207]

Is the alleviation of poverty just a fantasy? Whatever the ultimate prospect, and however the end point is defined, the process will necessarily be a long one. It involves eradicating corruption in government, building the institutional structures that are prerequisites for prosperity and creating the well-regulated markets needed to build sustainable wealth. In the poorest countries people subsist on the equivalent of just US$200 per year. If we assume that we can increase their income at the rate of 10 per cent per annum (which is extraordinarily ambitious), then by 2050 the incomes of the poorest will still be far behind those of developed countries.[208]

Even by the most optimistic of assessments, the world has a century of poverty ahead of it, but that does not mean that poverty-fuelled conflict is inevitable. Hope is a powerful tonic, even to those most sorely afflicted. In China the truth of this is evident—everywhere you look you see people who are happy because they now earn a dollar per day, whereas last year they earned just ninety cents. Upright apes are concerned with relative prosperity rather than absolute wealth, and it's a sense of relative improvement that helps keep the peace in the world's most populous country. And so it can do throughout the world, if we can keep improving the lot of the poorest.

The inhabitants of the developed countries use so much by way of resources that it's simply not possible for all of us to live that way without bankrupting the planet—indeed the rich are doing that right now all by themselves. It's clear that the affluent will need to reduce their consumption and to manage their expectations if they hope to protect their futures. But if raising the standard of living of the poor is challenging, reducing the consumption of resources by the rich is a far more difficult task. One way it may become more achievable is to propagate the right mnemes just as we have propogated the anti-smoking mneme. If we decry excessive consumption wherever we see it, whether in four-wheel drives on city roads or in oversized and energy-hungry houses, we may succeed. But this takes courage and individual action. All too often, when I see such things and want to say something, I remain silent for fear of social embarrassment.

One other useful mneme is the idea of such reductions in consumption allowing our children, as well as the poor, to live reasonable lives. After all, only a limited number of tonnes of CO_2 can go into the atmosphere before disaster happens. Only a limited number of trees can be turned into lumber before our world is deforested. Buying a smaller, more efficient car and driving it less often isn't much of a sacrifice; nor is eating sustainably produced food, or living in a more modest or more energy-efficient house.

The problem is that conspicuous consumption is tied up with our image of ourselves and one of our most ingrained social instincts—to keep up with the Joneses. To compete is fundamental to human nature, but using consumption to communicate our place in the race may be a fatal flaw. It is possible to devise a different form of social competition, one that doesn't rate big cars or big houses as signs of success but rather as emblems of ignorance and selfishness. Today, some of our wealthiest citizens follow such a path, and by giving their wealth away seek to build a better world. Not enough, however, take to heart the Scottish-American industrialist and philanthropist Andrew Carnegie's maxim that 'to die rich is to die disgraced'.

As yet we've barely begun to imagine how we might guide our society away from dangerous over-consumption by fostering prestige in different ways. Nor have we thought enough about how we might turn today's status symbols into emblems of ridicule or disgust. Instead, some have sought to ignore the world's problems by cutting themselves off from the world in gated estates. As the century progresses, the folly of this approach will become increasingly self-evident. There's no easy way of achieving a more globally equitable, sustainable world, but as long ago as 1961 US president John F. Kennedy made a start, opining that 'if a free society cannot help the many who are poor, it cannot save the few who are rich'.

There are other triggers to conflict in our modern age, the most important of which have roots in the dying of the tribal world. A globalised human culture, by virtue of its very nature, must subscribe to and support a humanitarian world view in which every person is awarded equal rights. This is resisted by adherents of religion or culture who believe that they are different from or superior to the rest of us. These relics of tribalism seek to separate people from each other, often on the basis of dietary restriction, cultural practice or dogma. To some extent it's understandable. Tribal ideologies can make us feel special—part of a chosen people—and to accept the beliefs or practices of others can be challenging. The roots of social dislocation, cult massacres and terrorism all lie here, and they're all likely to increase for some decades yet as globalisation exposes more and more people to their impact. This phase of human development is a kind of rocky road that must be traversed before we can reach a more stable state. Our hope must be that as the older generations pass, new generations born into the globalised world, with new ways of thinking about these challenges of war and poverty, will take their place.

CHAPTER 20

A New Tool Kit

Theatrum Orbis Terrarum
(Theatre of the World)
<small>TITLE OF THE FIRST TRUE ATLAS, BY
ABRAHAM ORTELIUS 1570</small>

The Forum of Vespasian in ancient Rome sheltered a marvel. Known as the *Forma Urbis Romae*, and carved in marble at the beginning of the third century, it was a map of Rome, then home to one and a half million. At a scale of 1:240 it covered an entire wall, and was considered so vital that access to it by anyone but authorised staff (who were equipped with long pointers) was forbidden. On it every street, wall and *popina* was marked with precision, and it served as the reference in all property disputes.

Without possession of a true, reactive and up-to-date map of the world, a sustainable future is impossible. Yet until recently we've had little better than the *Forma Urbis Romae*. When I was a child my window on the world was a set of old, leather-bound encyclopaedias that contained maps probably drafted in the early twentieth century—at least I vaguely remember the Austro-Hungarian empire featuring in them. While they showed some topographic features, their main characteristic was countries and their colonies marked in bold colours. I remember Australia and Canada being large splotches of pink, the colour of the British empire. Now, almost fifty years later, when I need maps I go to Google Earth. There I can see our planet in virtual real time and true colour, and so detailed is the coverage that I can even see my own little nest with its solar panels bright among the gum trees. On Google Earth the national boundaries have receded, replaced by our living planet in all its glory and complexity. And on any decent contemporary globe, even the depths of the oceans, which in my childhood were just a blank, are now mapped, their submarine ridges and ranges snaking across the blue.

Possession of the right kind of map is vital to solving our environmental problems, but there are other requirements. For a start, we must be capable influencing Earth's organs—its crust, oceans and atmosphere—in order to help maintain the planet's chemical and thermal balance. That's a high bar for a species just emerging from a tribal age, for it requires unanimity of purpose, informed by a profound scientific understanding, and intelligent and responsive tools of management. While we are far from attaining any of these, advances made in computer technology give hope that we are on the threshold of an extraordinary breakthrough.

A little over a decade ago I was the intelligence in the car I drove, as well as the physical force that directly controlled it. There was no assisted braking or steering in my old jalopy, no cruise control, computerisation of fuel and engine, nor a GPS. In fact there was no computer in the thing at all. Today, my hybrid car can park itself; indeed, it seems on the verge of being able to drive itself. And it's not only cars that are getting smart. Computers are increasingly involved in all forms of transportation, in the electricity grid, food production, waste disposal and water delivery, to name just a few areas being revolutionised by the silicon chip. And computerised systems are efficient, responsive and integrated—characteristics that, as our population nears seven billion, are essential to keep our demands for Earth's resources from breaking the Gaian bank.

A profound transformation, which is closer than many think, concerns the convergence of the transport and energy sectors. Plans are already underway in a number of countries for the large-scale deployment of electric cars. The most advanced plans result from a partnership between Denmark's largest energy provider, DONG Energy, and a US–Israeli start-up called Better Place. At the launch of the partnership I heard first hand of DONG's plans to spend hundreds of millions of kroner on electrifying car-parking spaces throughout Denmark, as well as in constructing battery exchange terminals that would permit car batteries to be replaced in less time than it takes to refuel a conventional vehicle. I asked DONG Energy CEO Anders Eldrup why his company was pursuing this initiative. He explained that the business had built up a tremendous portfolio of wind assets, and that these were underperforming in an economic sense because no one wanted the electricity they produced at night, which is when the wind often blows most strongly in Denmark. By charging electric cars at night, Anders saw a means of developing a market for the electricity.

Things get a little complicated when you suddenly make such large demands of the electricity grid. What would happen, for example, if everyone returned from work at the same time and tried to charge their vehicles? The grid might crash, putting vital services such as health care at risk. Clearly, a far more responsive and 'intelligent' electricity grid than Denmark currently has would be required. And that is where Better Place comes in. It is able to provide smart metering that permits the grid provider to charge electric vehicles at times and at volumes that avoid collapse and maximise the use of renewable energy. It also simplifies billing, providing monthly accounts just like those for mobile phones. Eventually Better Place hopes to be able to draw electricity from cars that are plugged in to assist with load management at times of high demand.

It's clear that the uptake of electric vehicles will drive the adoption of smart-grid technologies, maximise the efficiency of electricity use, minimise the need for new electricity generation and encourage the uptake of renewable energy sources like wind and solar. When we consider the scale of city building in the developing world—where three new United States' worth of cities will spring up in the next forty years—it's evident that conventional technologies won't be able to deliver the transport, energy supply and air quality these cities will require. Electric vehicles, along with investment in public transport, offer the best solution, and I have not the slightest doubt that they will dominate our motoring future.

The smart grid, of course, offers benefits beyond fueling electric cars. Our existing electricity grid is about as dumb as it gets. Demand drives its dynamics, and there's little an energy provider can do other than blacking out large sections of a city to prevent overall collapse when, for example, huge numbers of people turn on air conditioners on a hot afternoon. But imagine a grid that would allow the generator remotely to switch off refrigerators for twenty minutes at a time, or pool pumps, or some other non-essential item, at times of high demand. Or imagine a grid that delivers control to the consumer—a smart grid that can efficiently utilise the electricity generated by rooftop solar panels. The difference between a smart grid and our existing model is about as great as that between my old jalopy and the hybrid car, only it doesn't have to be more expensive, for the efficiency savings promise to offset the initial costs.

In future the smart grid may resemble our autonomic nervous system—the portion of our nervous system that is not under conscious control but which regulates our breathing, digestion and heart rate. Our autonomic nervous system makes us efficient through the subliminal, whole-of-body coordination of vital functions; and so the smart grid could, through subliminal control of a city, promote hitherto unimagined efficiency. Such quantum leaps in technology require prodigious investments. Imagine replacing all the petrol stations, electrical infrastructure and motor vehicles in your country. And these investments cannot be disaggregated, for one is dependent upon the existence of another to succeed. The projected cost for smart-grid infrastructure in the US alone over the coming decade is estimated to be around US $400 billion, and it's yet to be seen how this would be financed.[209]

What of technologies that will allow us to apply intelligence to other human activities that affect Gaia? A good place to begin is with the most intimate interface between ecosystems and human societies—agriculture. Anyone living in a dry region of the world will be aware of the revolution that's occurred in irrigation technology in recent decades. In my childhood, water was applied to crops in ways that had not changed since the time of the Chaldeans. On my uncle's dairy farm precious water was transported in leaky channels, and fields were simply flooded. Today, in the same region, pipelines and computer-operated drip-feed systems deliver the precise amount of water required by plants for optimum growth. Yields are increased and water use decreased by an order of magnitude at least.

Such changes are important because about 12 per cent of Earth's surface is intensively used agricultural land, and its management is critical to optimising stability and sustainability in Gaia. The first significant congress on the use of computers in agriculture was held as recently as 2003. It was a modest, mostly North American affair, but by the time of the 7th World Congress on Computers in Agriculture in Nevada in June 2009, the number of topics and countries represented was astonishing. From wireless monitoring of microclimates in fields to moisture availability and automatic detection of lameness in dairy cattle, and on to robotic laser weeding systems and automated optical assessment of fruit, it seemed as if every aspect of agriculture was being made more efficient through the use of computers.[210]

Of course, the deployment of computers in agriculture won't of itself cure the Earth's ills. It must be coupled with a detailed understanding of ecosystem function and impacts on the carbon and nutrient cycles. But the point is that we now have tools that allow us to relate to living ecosystems with an efficiency and speed of response that we have never before possessed. Is it possible that we might one day regulate non-agricultural systems in a similar manner? The first steps in that direction are already being taken, through the remote surveillance of forests. In January 2009 the government of Malaysia announced that it would use satellite monitoring to ensure that illegal logging 'is stopped with immediate effect'. Around one-third of Malaysia's annual deforestation is carried out illegally, so this represents a significant advance.[211] In Brazil, the benefit of satellite surveillance is already paying off. The amount of deforestation between August 2009 and May 2010 was only half that of the same period the previous year, a decline which Luciano Evaristo, the country's director of environmental protection, attributes to improved satellite surveillance: 'Before we were looking blindly. But in 2010, all 244 actions were based on smart geo-processing.'[212] On a broader scale, the satellite monitoring of forests is already capable of exposing all illegal logging worldwide.

Satellite monitoring has existed only for a few decades, yet it's revolutionising our understanding of how Earth works. Among the most ingenious of all monitoring systems are the paired satellites that allow detection of gravity anomalies that can be caused by ice-melt. With these tools, it has become possible in just the last few years to measure with precision changes in such remote regions as the Greenland and Antarctic icecaps. We are already capable of detailed monitoring of Earth's atmosphere, and are slowly working our way towards a global monitoring system for the oceans—which is a much more difficult task because the ocean is opaque and five hundred times larger (in mass) than the atmosphere. Beginning in 2003, a multinational project allowed three thousand Argo probes—which can descend to a depth of two kilometres and resurface with recorded information—to be lowered into the oceans, but ten times as many are required to provide a sufficient level of detail to reveal what is really going on in the depths. At present some of the gaps are being filled by fitting large sea creatures, seals, sharks and other animals, with monitors that send data to recording stations.

I can imagine the day when our surveillance of the atmosphere, the oceans, the land and the heavens is so complete that we will be able to anticipate most natural disasters. Such a system would also give us fair warning of when human intervention is required to disperse malign trends. I can also imagine a time when that 12 per cent of the land surface that is used for intensive agriculture will be managed so that carbon flux and productivity can be controlled effectively, and when the rest of the Earth's land surface is carefully monitored for destructive changes. We have or are rapidly gaining the tools needed to do this, but the political agreements needed to use them wisely are yet to be realised.

CHAPTER 21

Governance

*The common world has been the work
of every generation that has lived
in it, back to the remotest ages.*
JONATHAN SCHELL 1982

Even monkeys play at politics, so we can be certain that for as long as there have been people there have been politicians. Today a diversity of governments exists, from relict monarchies such as those in Tonga and Saudi Arabia to theocracies in the Vatican and Iran, and to harsh dictatorships such as those in Myanmar and North Korea. There are also monopolistic governments which are flirting with democratic representation at some level, such as in China and Russia. But for all of that, as Francis Fukuyama wrote in *The End of History and the Last Man*, humanity has increasingly settled upon one political system, which in time looks set to supplant all others.[213] Fifty years ago there were only forty or so democratic countries on Earth. As I write there are 123. The prospect of dictators flourishing in western Europe now seems remote, but I grew up in a world of Francos and Salazars, and they've only been gone thirty-odd years. The spread of democracy during the second half of the twentieth century is surely one of the most remarkable political phenomena in human history.

The growth of democracy is vital to a sustainable future. It alone can provide security and secure rights, such as property rights to the individual, which ensure that most have 'something to lose' and so will not steeply discount their futures. While the number and strength of democracies are growing, we still have a very long way to go before we live in an entirely democratic world.

From a biological perspective, the motivating force of democracy is little more than a continuation of the trend that began with the dawn of agriculture, wherein the individually weak farmers triumphed over the powerful few. Within democracies the struggle between the people and the powerful few continues. From New York to Zurich a privileged minority still exerts disproportionate influence. How we create better democracies to deal with this is a defining challenge for humanity. I find it astonishing that in most democracies it's still legal for anyone to make payments to political representatives, and by so doing subvert the interests of the voters. In a true democracy all political funding would come from the people as a whole, not just a select few, for whoever pays the piper calls the tune. A landmark event occurred in 2008 with the election of US president Barack Obama. By using the internet to gather small donations from a large number of relatively powerless people to fund his campaign, Obama changed the democratic process.

The UN, with its mix of democratic and despotic governments, remains the closest thing we have to world governance. Entering the chamber of the UN General Assembly in New York is like being teleported back to the 1950s. Just as on my old map in the encyclopaedia, it's countries that count here. Each nation has its own desk. The desks are uniform in size and arranged alphabetically. Only two—the observer states of the Vatican and Palestine—are out of order, sitting at the back of the room. Weighty matters are debated in the chamber, and in this strange world in miniature the voice of the peoples of San Marino (population thirty thousand) and Monaco (thirty-three thousand) each carry as much weight as the billion-strong throngs of India and China and the economically powerful USA and Germany. As a result it's possible for a binding vote to be carried in the UN General Assembly by a two-thirds majority, yet be supported only by representatives of 8 per cent of the world's population.

Perhaps places like the UN General Assembly and the European Parliament tell us that all high levels of government, being so far removed from the people, are more likely to become irrelevant and ineffective. But is a strong world government necessary for a sustainable human future? Several examples from the natural world suggest that this may not be so. Ants, after all, maintain their large and highly complex societies without the help of a 'brain caste', or indeed any kind of blueprint for intelligent management. And our own brains illuminate how a superbly complex and competent command-and-control system can be made up of rather poorly integrated parts.

Our brains consist of three broad areas of responsibility. The two hemispheres (what we normally think of as 'the brain') are significantly enlarged in humans and linked principally through a bundle of nerve fibres known as the corpus callosum. They're responsible for rational thought, language and other high functions. Below them lies the old mammalian brain, composed of the amygdala, hypothalamus and hippocampus. Its structure in humans differs little from that in other mammals, and it's vital to the emotions and long-term memory. The most ancient part is the so-called reptilian brain, located below the rest and made up of the brain stem and cerebellum. It controls instinctual survival behaviours, our muscles and the autonomous nervous system.

There is no 'brain commander' that controls all activities within this composite brain of ours. As we are all aware, on occasion conflicts arise—between emotions and rational thought, for example—that result from this layered evolutionary history. Yet none of this prevents our brain from functioning as an effective command-and-control system most of the time. Indeed, even when the connections between various parts of our brain are damaged, we still manage to function, as can be seen in people who have had their corpus callosum severed in an attempt to control epilepsy. In some people so treated the two independent hemispheres begin to act quite separately—the hand controlled by one hemisphere, for example, will undo something that the other hand has just done. But the majority of people continue to function fairly normally, perhaps partly because communication between the hemispheres is achieved externally through vision or sound.

All of this is to say that an effective governance system need not be ruthlessly centralised, but merely capable of sending messages that effectively influence the system it seeks to control. And this makes me think that our global superorganism can function perfectly well without a single strong, centralised government. But before jumping to that conclusion, let's look at how we are currently managing those parts of Earth that lie outside national boundaries and therefore require a coordinated global approach—the atmosphere, the oceans and the poles.

As I consider the struggle to manage our atmospheric commons, I cannot help but write from a personal perspective. In 2007, two years after publishing *The Weather Makers*—a book that outlined the science behind climate change—I more or less left scientific research behind to help establish a business council with global reach and experience whose aim was to amplify the voice of progressive corporate leaders on the climate issue and to build momentum towards the UN summit in Copenhagen.[214] Known as the Copenhagen Climate Council, it brought together the best of science and policy advice with the CEOs of leading companies. In May 2009 we held the World Business Summit on Climate Change, which attracted eight hundred business leaders. In September of that year CEOs from our council participated in the UN climate week, leading roundtable discussions with about eighty heads of government, almost half the world's political leaders, on aspects of climate change.

I'd never felt as optimistic about the prospects of humanity overcoming its greatest challenge as I did in September 2009. The sense of collegiality among heads of government was palpable, and the message that business wanted an effective deal struck at Copenhagen was delivered loud and clear. A few days later, when the G20 met in Pittsburgh, the leaders in attendance stated that they would 'spare no effort to reach agreement in Copenhagen'. But just weeks later, things began to go backwards when it became clear that the US Senate would not pass its climate legislation in time for the Copenhagen meeting. It felt as if the world was being held to ransom by a few hold-outs whose reflex beliefs in the 'survival of the fittest' permitted violations of the common good.

The final blow to my hopes for a comprehensive deal came on 24 October 2009, the day when thousands joined in global demonstrations to support a target for the atmospheric concentration of CO_2 of 350 parts per million. As they gathered on the plains of Africa and Central Asia, on the Great Barrier Reef and in Washington, news was leaking out that Yvo de Boer, the executive secretary of the UN Framework Convention on Climate Change, the entity that organises the global negotiations, was saying that no legally enforceable treaty would be agreed at Copenhagen. It felt as if, having swum against the pull of the whirlpool for so long, we were in danger of being sucked into a vortex of unbearable bleakness.

But what of the meeting itself, which was held in Copenhagen's Bella Centre? I wrote the following paragraph late at night, flying home from the summit:

> From the start flaws were evident. The meeting was poorly organised by the UN, but that was only part of the problem. The Danish prime minister Lars Rasmussen was inexperienced in global politics. Sudan was head of the G77 (a grouping of the less developed countries) and when Amnesty International called on Denmark to arrest President al-Bashir of Sudan over breaches of human rights, Sudan set out to wreck the negotiations.

On the last day of the summit, Friday 18 December, after weeks of fruitless meetings and uncertainty, exhaustion had taken hold. I left on the Saturday afternoon—with the negotiations still continuing long beyond their scheduled termination time of midnight on Friday—with a sense that something fundamental had shifted. The UN process that had guided climate negotiations for fifteen years had proved itself, in the end, incapable of delivering a decision. It was only through President Obama breaking into a meeting with Chinese, Indian, South African and Brazilian leaders that a deal of any sort was struck. And even that deal, the Copenhagen Accord, was not adopted by the UN, but merely 'noted'. It's hard now to imagine the UN delivering any global deal on any matter of major significance. Whatever governance model humanity settles upon to resolve its problems, it's unlikely to be via a convocation of nations united as equals.

The Copenhagen Accord may appear to be a weak outcome,

but that is largely because our expectations were so high. It has been dismissed as a merely political agreement, but any agreement that includes the largest carbon emitters should not be sniffed at. The developing nations own the outcome, so they can no longer shrug off their own responsibility to act on climate change. The reduction pledges made by China are so large that the technocrats consulted on them replied that they were not feasible. But the Communist Party persisted, and they are now being implemented as well, incidentally, as putting in place a cap-and-trade scheme for greenhouse gases. If the reductions agreed to are in fact globally realised by the 2020 deadline, they have the potential, if developed countries play their part, of seeing atmospheric CO_2 concentrations stabilise at around 450 parts per million, or 2° Celsius of warming. If all carbon reduction pledges under the Copenhagen Accord are met, humanity will be emitting just forty-eight gigatonnes of CO_2 in 2020. To stay below 450 parts per million we'll need to be emitting forty-four gigatonnes, so we're just four gigatonnes short. And that could be made up if developed countries cut their greenhouse-gas production by about a third more than currently pledged.

Some worry that the Copenhagen Accord is not a legally binding treaty, but can any treaty be truly binding within a UN context? Under the legally binding Kyoto Protocol we've seen countries such as Canada repudiate its obligations without consequences—indeed it's difficult to see how any treaty could be enforced. All that can be said right now is that it's too early to declare the Copenhagen meeting a failure.

Perhaps we shouldn't be surprised by the difficulties the world had at agreeing in Copenhagen. In 2008 game theorist Manfred Milinski published the results of a simulation of the Copenhagen summit. One hundred and fifty-six student volunteers were divided into twenty-six teams. Each participant was given a budget of forty euros and was told that unless they could stop climate change a human catastrophe would occur. In order to be effective in their fight against climate change, however, each team of six people would need to contribute 120 euros—an average of twenty euros per person. So, in the game, there was enough money to solve the problem. Over ten rounds each student got various chances to contribute a small amount, a large amount or nothing at all. If their team succeeded in meeting the climate challenge individuals got to keep any leftover cash. But if they failed they got nothing. In a scenario where climate doom was certain, only half of the teams contributed the 120 euros required to succeed, while in a version of the game where the chance of catastrophic climate change was reduced to 10 per cent, only one team out of twenty-six contributed sufficient funds.[215]

Clearly the students were not acting rationally or in their best interests. One reason they might have done so poorly is that they had no way of punishing those who didn't contribute towards saving the climate or of rewarding those who did, which sounds a lot like the current UN process. Commenting on the prospect of success at Copenhagen prior to the event, the veteran game theorist Carlo Carraro said 'there's no chance'. Any deal with a chance of success, he believes, must focus on carrots and sticks—that is, incentives for countries to reduce emissions and disincentives to cheat—and those incentives must be of the right type.[216] Game theory indicates that carrots are several times more effective than sticks, and that the worst kinds of sticks are unilaterally imposed ones, such as the trade tariffs for carbon-intense goods included in the US's proposed Waxman-Markey legislation, which are more likely to enrage nations than provide incentives for cooperative behaviour. This bill was passed in the House of Representatives in 2009, but has yet to be considered by the Senate.

By highlighting the inadequacies of the UN process, the Copenhagen summit invited reflection on how we might deal with global problems in future. One possibility is that the G8 (France, Germany, Italy, Japan, UK, USA, Canada and Russia), the G20 (a group of twenty finance ministers and central bank governors representing the largest economies whose meetings heads of state have recently been attending) or the Major Economies Forum (representing seventeen major economies, many with high per-capita carbon emissions) might provide a workable approach to finding a solution. From a game theory perspective this is preferable because the possibility of brokering an effective deal is greater if a small number of players is involved. If we learn from our mistakes, our chances of succeeding at our next attempt will improve.

The real world of global climate politics is infinitely more complex than games. Senates, or houses of review, in government systems can impede progress, and so can the UN-based system of one vote per nation regardless of population. In both instances, the problem lies in the reluctance of political entities to cede power. Perhaps a detailed study of successful and failed federations, as well as successful and failed global treaties, can inform progress here. Another problem for the UN lies in the prevalence of dictatorships and partial democracies. Being devoted to the good of a few rather than all, they act as corrupting elements whenever the state power of democracies is ceded by treaty or other means.

Copenhagen was not only about the political negotiations. One of the more extraordinary initiatives of the meeting occurred outside the political mainstream. The Vote Earth campaign is a partnership between WWF's Earth Hour and Google Earth. It distributed electronic ballot boxes across thousands of web portals, inviting people to 'vote Earth' in support of a strong outcome at Copenhagen. While it failed to deliver its strategic objective, the campaign made extraordinary advances in web-based democracy. With protocols in place to limit each web address to one vote only, the system goes some way towards ensuring one vote per person, something that just two years earlier the internet was unable to deliver.

With 3.3 billion active mobile phones, and many people able to access the internet via internet cafes or their own computers, the Vote Earth campaign demonstrates the possibility of globally participative democracy that, to some extent, circumvents the power of the state. I can imagine, for example, a group of citizens using such tools to organise a vote for the position of a de facto secretary general of the UN. Such an individual may not have formal power, but he or she would surely have great moral suasion. Perhaps such online elections, organised by the people, of the people and for the people, may in future be held in parallel with official elections. Just how our societies will cope with the power of the internet to focus mass political will remains to be seen, but it does represent a means of redressing, at least in part, the undemocratic tendencies inherent in many governments.

We must now turn to the management of our other global commons. A novel development is currently underway in the Arctic, whereby nations surrounding the icecap are quickly appropriating as much of the north polar commons as the UN Convention on the Law of the Sea (UNCLOS) allows. Under that treaty, a nation wishing to claim part of the ocean needs to demonstrate that the seabed is an extension of its continental landmass. As of mid-2009 the only unclaimed parts of the Arctic were a wedge of seabed on the Russian side of the North Pole and a couple of small patches north of Alaska. Whether this appropriation of a global commons is any more successful in fostering sustainable management and use, however, is yet to be determined.

Antarctica offers yet another means of dealing with a global commons. Seven nations claim sovereignty over parts of the ice-covered continent. Under the Antarctic Treaty, which has been ratified by forty-seven countries, the continent is set aside as a scientific preserve and there is a ban on military activity. In 1983, the Seventh Conference of the Heads of State or Government of the Non-aligned Countries launched a direct challenge to the treaty. They'd missed out on staking a claim and felt that exploitation of Antarctica's natural resources should be carried out for the benefit of all humankind, arguing for 'widening international cooperation in the area'. While extraction of resources from the Antarctic continent may be some decades off, the harvesting of fish and other resources from its surrounding seas is already occurring, and the push is on for declaring the entire Southern Ocean *res communis*— part of the common heritage of humankind—which would subject fishing to international taxation, user charges and leases, or the sale of fishing permits. But how well are the oceans being managed under the rule of *res communis*?

It was Dutch jurist Hugo Grotius who, in the seventeenth century, developed the concept of the freedom of the 'high seas', stating that the open ocean was international territory that anyone was free to use. In 1967 Malta's UN ambassador Arvid Pardo argued before the UN General Assembly that the seabed on the high seas should also be considered the 'common heritage' of humankind. The speech sparked a chain of events that, in 1994, saw the UNCLOS come into effect. Among its most important provisions was the establishment of a UN agency, the International Seabed Authority, to regulate the extraction of mineral resources in the high seas. It also saw an extension of the rights of nations to the seabed from the traditional three-mile limit (4.8 kilometres) to twelve miles (19.2 kilometres) from the coast, and the granting of exclusive

economic use of a zone extending two hundred miles (320 kilometres) offshore. While these changes have seen most fisheries fall
under national control, there are still vast areas where *res communis*
rules, and great hunks of Earth's biodiversity and possible mineral
wealth reside there, awaiting exploitation by the first comer.

No example of our utter failure to manage the *res communis*
is more dismaying than that of the Atlantic bluefin tuna. True
denizens of the blue ocean, they can grow to three-quarters of a
tonne in weight and, swimming at up to a hundred kilometres per
hour, they cover huge distances, using some national jurisdictions
to breed in and others to feed in, and are equally dependent upon
all. In order to conserve this valuable resource, some forty years ago
a managing body was set up—the International Commission for
the Conservation of Atlantic Tunas. It represents forty-three nations
and is mandated to conserve the fish, but it's been controlled by
national fishing organisations, and has done nothing but accelerate
the tuna's demise. As American marine biologist Carl Safina noted,
Atlantic bluefin tuna are just too valuable to let live anywhere. One
201-kilogram fish sold wholesale in Japan for US$173,600, making
it the most valuable wild creature on Earth, worth more than a
poached elephant or rhino. As 2008 drew to a close, Safina wrote:

> End-times loom for the giant bluefin tuna, whose chances
> of survival were greatly diminished in late November by
> the international commission charged with its care. Once
> again, that body…refused to take strong action to prevent
> the runaway overfishing of the giant bluefin tuna in its
> sole remaining, yet rapidly disappearing, stronghold: the
> Mediterranean. [By setting the 2009 Mediterranean catch
> limit at twenty-two thousand tonnes,] by incompetence,
> greed, and reckless industry interference…the commission-
> ers agreed to ensure further decline.[217]

Fisheries experts consider the species to be on the brink of extinction, and the commission charged with its preservation has barely restricted its catch.

It was American ecologist Garrett Hardin who, in 1968, awoke us to the tragedy of the commons.[218] But Elinor Ostrom considers it is possible to manage them sustainably if we have the ability to exclude outsiders; clear, mutually agreed rules about who is entitled to do what along with appropriate penalties for transgressors; an ability to monitor the resource; and mechanisms to resolve conflict.[219] Until very recently we have lacked, and indeed in some cases we continue to lack, these abilities in the context of our global commons. It's worth noting, however, that human surveillance is now pretty much capable of detecting transgressors and monitoring resources globally. What we still lack are the political aspects—the clear, agreed rules, the penalties and the conflict resolution—and until we achieve them through the only feasible mechanism we currently possess, which is a global treaty, we are likely to continue to fail to manage our global commons, to the detriment of all.

Ultimately, such a treaty must take a holistic approach that protects the chemistry and ecology of these places. Perhaps it will be administered through a future Gaian Security Council, sufficiently empowered and constituted so that it is the ultimate authority over global commons. It's hard to imagine the nascence of such an organisation today, but the children of a globalised world, looking back at our abject failures, might give it birth. The trouble is that the threats to our environment just keep growing and now have the potential to overwhelm us before we achieve such wisdom.

A ray of hope, however, can be seen in quite a different direction. The destruction of our global commons thrives on secrecy, and in the twenty-first century, courtesy of globalisation and new technologies, that is becoming a rare commodity. It's possible to create a website where anyone who sees pollution or overexploitation occurring could upload that information into a globally accessible database, along with pictures from mobile phones, satellite images and laboratory reports. Imagine a detailed database of the world's coal-fired power plants. Imagine the same for the overexploiting fishing fleets, or the dumpers of toxic waste. If moral suasion has any power at all, it may be possible that the good citizen, thus informed, could buy us the time to develop the international approach required to establish sustainable management of the global commons. And the citizenry may not stop at moral suasion. The case of Greenpeace taking direct action against the Japanese whaling fleet in the Southern Ocean, using their boats to get between whales and the whalers, is just one example of people, fed up with the inabilities of governments, taking matters into their own hands on behalf of the planet.

CHAPTER 22

Restoring the Life-force

There were some also [who] held, that if the
Spirit of Man doe give a fit touch to the Spirit
of the World…it might command Nature.
FRANCIS BACON 1639

Can we expand Earth's biocapacity, its potential to sustain life? On a planet being asked to support nine billion people nothing is more essential. It's Earth's biodiversity that keeps it habitable. The first requirement is a dramatic reduction of greenhouse-gas emissions. But that alone will not ensure our success, for the Earth system has been damaged by our activities, and that damage must be repaired. But how much damage has been done?

Carbon provides a way to calculate this. Over the last three centuries, by slashing, burning and ploughing, we have released between 200 and 250 billion tonnes of carbon. That's a quarter of a trillion tonnes of carbon effectively 'put to death' and it makes up between 22 and 43 per cent of all the carbon released into the atmosphere in that period.[220] Within that tonnage are entire species, from thylacines to dodos to rainforest trees, whose loss inevitably weakens Earth's resilience, as well as its productivity.

As yet we're unable to say by how much this loss has decreased Earth's energy budget. But it's indisputable that the destruction has impacted Gaia in two fundamental ways: by impairing Earth's carbon pump, which continuously turns atmospheric CO_2 into living things, and by supercharging our atmosphere with dead carbon, thereby putting pressure on the planet's climate system by warming the atmosphere. If we are ever to repair this damage, we'll need to develop a holistic view of the carbon cycle that allows us to apply 'a fit touch' to those Gaian tissues most critically damaged, as Francis Bacon put it in *Sylva Sylvarum*.[221]

Photosynthesis stands at the heart of Earth's productivity. Described scientifically, it sounds mundane—a plant takes in water, sunlight and CO_2 and gives out oxygen, at the same time using the carbon from the CO_2 to make sugars, which build the plant's tissues. A leaf is a small miracle, for through it a transubstantiation occurs—of a lifeless gas into a solid, living being. It's a sort of resurrection of CO_2, the gas given off with death and decay, the gas that enshrouds dead planets. Yet from it plants forge beauteous forms that support all the hosts of earthly life, ourselves included. We commonly misunderstand how trees grow, imagining that they somehow spring from the earth, from their roots. But this is not the case. Trees grow from the air, through tiny holes in their leaves called stomata, into which they draw CO_2. Look at a plant and you can roughly estimate (if you can imagine its roots) the amount of carbon it has sequestered during its lifetime. Cut down and burn that tree, and its lifetime store of carbon is released into the atmosphere.

While industrial methods of carbon capture remain on the drawing board, plants are the most effective mechanism in existence, breaking down 8 per cent of all atmospheric CO_2 each year. What we must aspire to now is storing some of that captured carbon. There's no better place to start than in global agriculture and forestry, for it is there that we are most intimately connected with the miracle that is the transubstantiation of carbon.

Life's carbon stores are largest in the great belt of tropical forest that girdles the Earth. Although they cover just a small percentage of Earth's surface, the rainforests are disproportionately high in importance to Gaia's climate system. In addition to storing carbon they are rain makers (four-fifths of the Amazon's rainfall comes from the forest itself, through transpiration of water vapour), and the forests' transpiration acts to cool our globe. They are also major stores of biodiversity, with an estimated two-thirds of all living species residing in them, and are home to hundreds of millions of Earth's poorest and most underprivileged citizens, who currently depend for survival upon some of humanity's least sustainable practices.

The potential of tropical forests to contribute to solving the climate crisis was examined in the Eliasch Review, named after the London-based Swedish businessman who was UK prime minister Gordon Brown's special representative on deforestation. Johan Eliasch spent £8 million buying 1600 square kilometres of Brazilian rainforest to protect it. His 2008 report was based largely on the figures supplied by the IPCC, and while unduly conservative in its estimations, it's a good starting point. It reports that on average one hectare of tropical forest takes up approximately three tonnes of CO_2 per year. Without this contribution to atmospheric cleansing, the report estimates, the concentration of CO_2 in the atmosphere would have risen 10 per cent faster than it has over the last two hundred years. But it has recently been shown that the tropical forests are in fact taking in 20 per cent more carbon than the report estimated, and so those tropical trees that remain are growing larger than ever before.[222]

Hitherto we have not valued the contribution of the tropical forests to climatic stability. Instead we have hewed and felled them, turning them into exotic timbers, such as teak, ebony and meranti, with which to build and adorn our homes. Yet, for all our axe-work, until the nineteenth century the tropical forests had survived pretty well. When Alfred Russel Wallace visited the island of Singapore in 1862, he found the place covered in dense and venerable jungle, in which tigers were still abundant enough to 'kill on average a Chinaman every day'.[223] Not long before that, in the seventeenth century, Javan rhinoceros, tigers and leopards lurked just beyond the stockade of the Dutch settlers at what is now Jakarta, and Hong Kong was a quiet fishing village.

By 2009, around half of the tropical forests present in 1800 had been destroyed and, at the current rate of destruction, by 2050 most of the remainder outside protected areas will be gone as well. As a result, some developing countries have high rates of greenhouse-gas emissions. Papua New Guinea, for example, produces a third as much greenhouse gas as Australia, a country with four times the population and which burns huge quantities of coal. Globally, 15 per cent of all human-caused greenhouse-gas emissions result from the destruction of rainforests. If we could reverse this dismal trend, and by 2050 restore between 8 and 17 per cent of what we have destroyed, then between forty billion and two hundred billion tonnes of CO_2 could be sequestered in the growing rainforest. Given that we've put around two hundred billion tonnes of CO_2 into the atmosphere over the past two hundred years, that reversal could theoretically get close to balancing Gaia's carbon books.

The destruction of the tropical forests is in the interests of very few people. Only the logging companies—who appropriate the patrimony of the original inhabitants, indeed the patrimony of all humankind—and the corrupt politicians who aid them benefit directly. Villagers who live in the region, in contrast, lose a source of building materials, food and medicines, while all of humanity loses a vital opportunity to stabilise our climate. But how is the situation to be turned around? Certification of forest products to ensure that they have been harvested sustainably is a powerful tool, but our approach to this to date has been voluntary and far too lackadaisical. A more concerted and holistic effort is required, and that can come about most effectively as a result of a global treaty.

If truth be told, we citizens of the developed world have been promising for years at environment meetings to pay to protect the world's tropical forests. First at the Rio Earth Summit in 1992, then at Kyoto, and again in Copenhagen. The poorest countries are bitterly cynical that progress can be made. Natural justice tells us that it must be done, and we can only hope that the US$100 billion promised by developed countries at Copenhagen will allow the poorest of our brethren to improve their lives in ways that protect the forests, and at the same time help obtain global climate security.

Tropical forests are not the only means of storing CO_2. Opportunities also exist in agriculture, forestry, rangelands management and even national parks. One technology, called pyrolysis, transforms biological carbon (the kind that's present in plants and animals) into a mineralised form. When plants and animals die they rot, releasing their carbon stores into the atmosphere. Mineralised carbon won't rot, so if it's added to soil it will stay there for hundreds or thousands of years. Pyrolysis works like a coal-fired power plant in reverse. Rather than feeding coal we've mined into a furnace and releasing its CO_2 to the atmosphere, pyrolysis uses the carbon captured by plants and turns it into a mineral form, which can be buried in the ground.

Crop waste, animal manure, forestry off-cuts, even human sewage, can be used as a feedstock for pyrolysis, and the process requires no external energy source except at start-up. The feedstock is heated in the absence of oxygen, separating it into solid, liquid and gas fractions. The solid is mostly charcoal (mineralised carbon), the liquid is a bio-oil, and the gas is made of carbon monoxide, methane and other compounds. Both the bio-oil and gas, which are rich in hydrogen, can be burned for energy. This releases some carbon into the atmosphere, but overall more carbon is removed from the atmosphere than is added. Alternatively, the bio-oil can substitute for crude oil in the manufacture of many products, from transport fuels to fertilisers and plastics.

Up to 35 per cent of the carbon present in the feedstock can be transformed into charcoal. If ploughed into soil, most of the carbon in the charcoal will remain there for hundreds or thousands of years.[224] Charcoal generated by pyrolysis is unique in that it is a long-term, safe and proven means of sequestering carbon. Most experts consider that a billion tonnes of carbon per year could be stored as charcoal in soils.

And there are other benefits. When dug into soil, charcoal decreases soil acidity and delivers residual nutrients and minerals. Bacteria and soil fungi essential to healthy plant growth soon colonise its porous structure. Its filtering capacity purifies water and assists moisture retention, enhancing plants' access to nutrients and moisture, thus providing a longer growing period. There are also indications that emissions of nitrous oxide, a powerful greenhouse gas generated by soil bacteria, are significantly reduced when soils are treated with charcoal.[225]

Numerous experiments indicate that yields across a variety of crops, from carrots to grains and fodder, generally increase when charcoal is added to soils. The impact is often greatest in leached tropical soils that lack carbon, with increases in yield of between 50 and 300 per cent recorded. In better quality soils yield increases, typically of around 7 to 20 per cent, have been documented.[226] Estimates of how much charcoal might boost global food yields are yet to be calculated. But anything that can help water quality and conservation, increase food production, produce clean energy and help fight climate change is a welcome addition to our basket of technologies.

The sequestration of carbon in tropical forests, or in the form of charcoal, is limited. Forests grow slowly—the optimal uptake of carbon by a newly planted seedling is decades away—and pyrolysis machines take time to build and get operating. This means that neither will be contributing optimally to combating climate change for a couple of decades. There are, however, other options that allow us to store carbon quickly and on a large scale. Principally, they involve modifications to the way we manage our agricultural soils and the world's rangelands (lands used for grazing), and the way we manage fire in the dry tropics, all of which can be broadly categorised as better management of our soils.

Soils represent a huge carbon reserve—around 150 billion tonnes worldwide, which is roughly twice the amount of carbon in the atmosphere.[227] It's three times as much as is contained in vegetation.[228] Soil carbon has three principal components: humus, charcoal and the roots and other underground parts of plants. Humus is a relatively stable, carbon-rich organic material composed of strong, long chains of carbon molecules. It's what makes soil look black. It has a large capacity to hold mineral particles, which are valuable to plants, and can absorb most of its weight in moisture. While it's an important element of soil carbon, humus is not the most prevalent form of carbon in our soils. That honour goes to living plant tissue, mainly in the form of plant roots.

The world's intensively used croplands have lost 30 to 75 per cent of their carbon content over the past two centuries; that is around seventy-eight billion tonnes of carbon. When combined with the carbon lost from poorly managed rangelands and from eroded soils (neither of which has been reliably estimated), it's clear that a huge amount of carbon has moved from soils into the atmosphere.[229] While this is bad news, there is a silver lining: with appropriate management it's possible to restore around two-thirds of the carbon lost from cropland soils within twenty-five to fifty years.[230] And for every tonne of soil carbon restored, 3.667 tonnes of CO_2 is drawn from the atmosphere.[231] (The apparent discrepancy is because the oxygen in the CO_2 molecule is stripped off during photosynthesis.) So, through restoring our intensively used agricultural soils, we could draw down around 140 billion tonnes of atmospheric CO_2.

Much soil carbon has been lost through traditional ploughing, which really is a declaration of war on biodiversity—the farmer rips out all life before planting a monoculture which is kept 'pure' with pesticides and herbicides. Modern ploughing practices such as 'zero till' and 'zero kill' involve the planting of a crop directly into pasture grasses. Such practices are creating a new agricultural revolution, one based upon coevolution's capacity to increase biological productivity and ecosystem stability.

Because they are hidden from us, it's easy to underestimate how voluminous plant roots are, and what an important job they do. The root mass of a tree is around the size of the above-ground growth, but for perennial grasses the root mass can be four times greater than the above-ground growth. While they are alive, plant roots add to the soil carbon by exuding more than two hundred carbon compounds, and when they die they add to the humus, both of which result in increased soil fertility.[232] The way we treat the above-ground bit of a plant has huge impacts on the roots. Research into root mass variability in grasslands in western China has shown that grazing intensity has a major influence on total root mass. The plant sacrifices root growth to keep replacing leaves lost to livestock, so that pastures with moderate grazing pressure have more root mass than heavily grazed pastures.[233] A root carbon gain equivalent to 3.3 tonnes of CO_2 per hectare, per year, could theoretically be gained by managers who switched from a severe to a moderate grazing regime.[234] But the benefits of the altered grazing regime are wider than that: the practice decreases erosion, enhances soil moisture retention and provides a healthier ecosystem.[235] It's been estimated that restoring degraded rangelands could sequester between about one and two billion tonnes of carbon per year.[236]

The world's rangelands are estimated to comprise more than 4.9 billion hectares, most of which is too dry, or the soils too poor, to support agriculture. By virtue of their large extent, their rapid response to changes in grazing regime and the relatively small numbers of people involved, it's possible that the world's rangelands offer the greatest potential to sequester large amounts of carbon in the shortest possible time. Large areas of the world's rangelands are not utilised by livestock, but instead form national parks and wildlife reserves. With the right fire regimes they could absorb large increases in soil carbon.

The Australian Wildlife Conservancy (AWC) is pioneering the strategic burning of small patches and ribbons of land early in the dry season in northern Australia as a means of preventing devastating wildfires. The program currently manages fire on about five million hectares of land in the central and northern Kimberley, including Aboriginal-owned lands, pastoral leases and conservation areas. In 2008 the program won the Western Australian government's top environment award. Minister for Environment and Youth Donna Faragher said, 'The prescribed burning program has made significant inroads into dramatically reducing the number and size of mid-to-late dry-season fires, significantly improving conservation management and protecting the region's biodiversity'.[237]

The method used by the AWC is loosely based on traditional Aboriginal burning practices, which have protected biodiversity and soils in Australia for thousands of years. The practices survived until recently, and the calamitous story of how they were disrupted has been eloquently told by a group of Western Australian researchers. They studied aerial photographs taken by the Australian military in 1953 as part of a project to develop a rocket range spanning the continent's northwestern desert regions. At the time the Pintupi people were living traditionally on their lands, undisturbed by the outside world, and their activities included the burning of small patches of vegetation. This created a mosaic of recently burnt, regrowing and maturing plots. By questioning Pintupi who later came out of the desert, the scientists confirmed that fire was used purposefully, frequently and regularly for many reasons, but mainly to acquire food by promoting the growth of edible plants and attracting animals, and to maintain many animal species. By the early 1970s, however, the Pintupi had left their traditional lands, and analysis of satellite imagery revealed that the patchy vegetation was soon being replaced with enormous expanses of uniform-aged vegetation resulting from widespread bushfires started by lightning strikes.[238]

Andrew Burbidge, who participated in this research, told me that, when he met the Pintupi in the 1980s, many were eager to see their country again, so he organised an expedition. Spirits were high as they set off, but as they approached their traditional lands the Pintupi became quiet. 'No one looking after the country,' an old man said. The Pintupi threw firesticks out of the car as they went, trying to breathe life back into the place, but when they reached a favourite campsite it took only a few minutes to confirm that all the middle-sized mammals were gone. The vast, hot fires that had occurred in their absence had destroyed the habitat (and taken much soil carbon with it). The group left completely dispirited.

Today Indigenous people are returning to traditional burning, and a computer study has shown just how much carbon could be saved continent-wide by such initiatives.[239] Six properties owned by Aboriginal people in tropical Australia were examined. The most promising was Hodgson Downs, a 3000-square-kilometre property in the Northern Territory. The carbon sequestered there was estimated to be around 10,400 tonnes of CO_2 equivalent (the warming potential, expressed in terms of CO_2, of all greenhouse gases involved) per year. Assuming a trading price of AUD$20 per tonne for carbon, in the event that Australia adopts an emissions trading scheme, income would amount to around $208,000 per annum.[240]

The potential to store carbon in rangelands used for grazing is also large. A grazing technique known as holistic management, which involves rotating stock and resting pasture, is revolution-ising rangelands management. It's being practised on approximately twelve million hectares worldwide, principally in Australia, Africa, Mexico, Canada and the USA.[241] Practitioners find that in addition to improving their pasture they can increase stocking rates by 50 per cent or more. A recent study of the rangelands of North American Great Plains concluded:

> Proper management of rangelands offers opportunities to partially mitigate the rise in atmospheric carbon dioxide concentrations through sequestration of this additional carbon via storage in biomass and soil organic matter.[242]

In other words, soil carbon is a potentially potent yet under-researched and underexploited contributor to returning Gaia to health.

Why are these practices so productive? Because they result in greater biodiversity and biomass in agricultural lands, allowing the win-win effect of coevolution. And they reduce farm costs. They replace a 'survival of the fittest' approach with a Gaian-based practice of ecosystem management, and in this they are a bright light shone on how we need to interact with the Earth as a whole.

Is it possible that such practices might forge a new, sustainable agricultural revolution, which could feed the projected nine billion mouths? The benefits are yet to be quantified at a global level. In Australia, however, in 2010 the Minister for Agriculture, Fisheries and Forestry Tony Bourke attributed increased grain and beef production, which has occurred over the past two decades despite a water crisis, to the uptake in 'zero till' or reduced-till lands and holistically managed herds.[243] If such practices are combined with a shift away from the feeding of grain to cattle in feedlots (a process that wastes 90 per cent of the energy in the grain), the application of intelligence to agriculture and the wise use of marine resources, I'm certain that Earth can support a future population of nine billion. It may not be able to do so indefinitely, but it might make it possible for us to pass humanity's population peak without catastrophe.

6

AN INTELLIGENT
EARTH?

CHAPTER 23

What Lies on the Other Side?

Do we as a species constitute a Gaian
nervous system and a brain?
JAMES LOVELOCK 1979

As I completed my lap of Charles Darwin's sand walk, I looked back towards Down House and wondered what the great man would have made of our world, with its cars parked over the fields where cattle once grazed, and his home turned into a scientific shrine. Would he have regretted that use of the term 'favoured races' in the title of his book? I'd like to think he understood that the survival of the fittest means the survival of none, and that he would have congratulated Wallace on his extraordinary insights, so far ahead of their time.

Had I just a moment with a resurrected Darwin, I'd like to share with him the spectacle of heavens' performance as we now understand it: from the moment of Earth's creation, through to the plains of Africa where our species took shape, and on to this century. We'd watch this Earth—a sphere of stupendous complexity—transforming itself over an immensity of time, guided by the evolutionary process that he so brilliantly elucidated.

If I were granted one conversation with Darwin, I'd ask him what he made of Bill Hamilton's last project. Hamilton, you might recall, was using computer models to investigate whether evolution builds ecosystems that, over time, become more resilient and stable: as he put it, how likely is it that a 'Genghis Khan species' will soon arrive and destroy all? It's a question that goes to the heart of this book, and I believe that the answer is to be found not only where Hamilton searched for it. It's a question that we, as individuals and as a global civilisation, must answer. Will ours be a Medean or a Gaian future? The choice will be made soon—for the best of our science and plain commonsense are telling us that our influence on Earth is eroding our future, and that we cannot escape responsibility.

If we take too small a view of what we are, and of our world, we will fail to reach our full potential. Instead we need a holistic, Wallacean understanding of how things are here on Earth with its illumination of how ecosystems, superorganisms and Gaia itself have been built through mutual interdependency. In this light it is absolutely clear that our future prosperity can be secured only by giving something away. But for the brief moment that is the early twenty-first century we strange forked creatures are perilously suspended between Medean and Gaian fates. Beckoning us towards destruction are our numbers, our dismantling of Earth's life-support system and especially our inability to unite in action to secure our common wealth.

Yet we should take solace from the fact that, from the very beginning, we have loved one another and lived in company, thereby, through giving up much, forging the greatest power on Earth. Those simple traits have allowed the weakest of us collectively to triumph, to establish agriculture, businesses and democracy, in the face of opposition sometimes so formidable as to make success look impossible. We have hated and fought too, but all the while villages have grown into towns, and towns into mega-cities, until at last a global superorganism has been formed. And today we understand ourselves, our societies and our world far better than ever before, and are uniquely empowered to shape our ends, rough-hew them as natural selection will.

Lately it's become fashionable to assume the worst, to imagine that our global civilisation has passed its peak and will soon collapse. Books like James Lovelock's *The Revenge of Gaia* and Jared Diamond's *Collapse* have done much to foster this philosophy, as has a growing awareness of the climate crisis. Cormac McCarthy's novel *The Road*, I think, captures the utter humbling of spirit that such an eventuality would bring. The moral aridity of that world, where life is reduced to a struggle for mere survival in a horrendous environment, is crushing. In it the division of bodies substitutes for the division of labour, and win-win is replaced by a catastrophic loss of all. We are capable of many things, but our beliefs do have a way of turning into self-fulfilling prophecies.

Climate science is now so advanced that we can anticipate the kind of event that may, if we do not reduce the stream of greenhouse-gas pollution, initiate the end of the great 'us' that is our global civilisation. With no warning, a gargantuan ice sheet will begin to collapse. It will mark the beginning of an irreversible process and, even if the initial rise in sea level it causes is just a few centimetres, it will herald an abandonment of our coasts, for the ice must continue to melt and collapse, albeit erratically, until there is no more. It will be impossible to put a time scale on the flooding, but Shanghai, London, New York and most other coastal cities must suffer partial or total abandonment, over weeks or decades or centuries. With economies in ruins and infrastructure drowned, we will then all be on *The Road*.

But the future worlds of Lovelock and McCarthy are just two possibilities. Such is the power of the mneme that, as far as our relationship with Earth goes, little is either possible or impossible—unless we think it so. Perhaps we'll tread a middle road, committing our global civilisation to a prolonged and agonising transition before securing a sustainable future. When a caterpillar weaves itself into its silken cocoon, it is for the most part weaving its own coffin. Shutting out the last of the light, it dissolves into a soup that nourishes just a few living cells that, by feeding upon the mush that was once their caterpillar, grow until they tear the silken veil to emerge as a moth and fly into the night. Must our human transformation be so brutal? With the climate crisis we're already sitting our first test, and it has arrived even before the human superorganism has properly matured.

But there is another possibility—that we will use our intelligence to avert catastrophe and secure a sustainable future. We now have most of the tools required to do this and, after ten thousand years of building ever larger political units, we stand just a few steps away from the global cooperation required. But do we have it in us to take those last steps? Between our evolved genes and our social structures, are we constituted so as to cooperate at a global level?

The immediate challenge is fundamental—to manage our atmospheric and oceanic global commons—and the unavoidable cost of success in this is that nations must cede real authority, as they do whenever they agree to act in common to secure the welfare of all. This does not mean the creation of a world government, simply the enforcement of common rules, for the common good.

By ordinary human measures, the climate crisis moves slowly, and so do the changes we're making to address it, so slowly, indeed, that we often fail to detect important thresholds except in retrospect. How will we know if we've turned the corner in our battle for a sustainable future? When profiteering at Gaia's expense is regarded and punished as the gravest of crimes—both because it represents a theft from the whole world, present and future, and because it may not remain mere theft but, as its consequences ramify, may become murder or genocide as well—then a sustainable future will be ours. Such a moment, if it ever comes, will close a chapter in human history—that of the frontier—which has characterised our species for fifty thousand years. In early 2010 we edged a fraction closer with the commencement of a campaign to have the UN's International Criminal Court recognise 'ecocide' (the heedless or deliberate destruction of the environment) as a fifth 'crime against peace'.[244]

If our civilisation does survive this century, I believe its future prospects will be profoundly enhanced, for this is the moment of our greatest peril. Should we cross the valley of death, democracy may well sweep the world, as Francis Fukuyama argued twenty years ago, creating a universal mode of government. And, as the geneticist Spencer Wells believes, in just a few generations most regional genetic difference will be muddied then lost. But there will also be tragic losses, for what is true of genes is also true of languages. Countless have already vanished, and the thinning of linguistic diversity will only continue as the members of our super-organism seek universal communication. Perhaps in the Chinglish of Singapore, Hong Kong and Shanghai we hear the embryo-genesis of a future world language. With a homogeneous gene pool, universal communication and a common political system, our children and grandchildren may have a far better chance than we do of acting as one.

It's sometimes argued that if humanity became extinct tomorrow, Gaia would look after herself. That may be true in the very long term—the tens of millions of years—but in the shorter term disaster would befall many species and ecosystems. That's because they've been so deeply compromised that only human effort keeps them functioning efficiently. The Australian Wildlife Conservancy, along with Aboriginal land holders, keep dozens of species from extinction. In New Zealand, and many other islands, species are kept in existence only through the most careful protection from introduced pests. Even in places like the UK, active management is required to preserve species such as heathland orchids and rare butterflies, while the majestic red kite soars in British skies only through our good grace. As the pace of climate change increases, our efforts to protect nature will become more critical.

This notion of humans as indispensible elements in the Earth system challenges the concept many of us have about our relationship with nature—for example, that we are somehow apart from it, or just one species among many. The truth is that no other species can perceive environmental problems or correct them, which means that the responsibility for managing this world of wounds we've created is uniquely ours. We are, it seems, the Faustus species—the one that, on that day thousands of years ago when we started to assemble our intelligent superorganism, signed a fateful bargain not with the devil, but with the blind watchmaker. It made us lords of creation, but left our fate and that of Earth inextricably interwoven.

As we seek to support the growing human family, our enormous power over nature could be exercised in any number of ways. We could, for example, seek to control nature at every turn, and so transform our planet into a huge, intensively managed farm. It's extremely doubtful, however, whether such an entity could be sustainably maintained, for it would lack the resilience and energy budget required to keep Earth habitable.

I would like to ponder a different kind of future relationship with our planet. The American ecologist Aldo Leopold said that:

> One of the penalties of an ecological education is that one lives alone in a world of wounds. Much of the damage inflicted on land is quite invisible to laymen. An ecologist must either harden his shell and make believe that the consequences of science are none of his business, or he must be the doctor who sees the marks of death in a community that believes itself well and does not want to be told otherwise.[245]

Such damage, we now know, stretches back over fifty thousand years, and it is profound. Making much of it good is beyond our current capacity, but, as an intergenerational ambition, healing Earth's ecological wounds is highly desirable.

There is something magnificent about the idea of a wild and free planet, one whose functioning is maintained principally by that commonwealth of virtue formed from all biodiversity. It's the sort of place celebrated by Jay Griffiths in her book *Wild*, which describes the last untamed corners of the Earth: places without roads, hotels and other western influences. Yet the book is as much a requiem as a celebration, for Griffiths acknowledges that wilderness is fast disappearing, if not already gone.[246]

If we wish to increase nature's influence a re-wilding is required—a reconstruction of vital ecosystems on a scale sufficient to allow them to operate optimally without intrusive human management. In effect, we'll need to make good the damage of fifty thousand years. Partially re-wilded areas in fact already exist in many parts of Europe, Africa and Australia, and in them one can see horses, elephants or wallabies roaming landscapes from which they were once long vanished. Such ecosystems are more productive and stable than the degraded ones they replaced. But

they are not as productive as they might be, and are far too small
to affect overall planetary health. The Russian biologist Sergey
Zimov has bigger plans. He wants to enclose part of northern
Siberia with a twenty-kilometre-long fence and introduce bison,
musk oxen, horses and other species long since locally extinct.[247]
This is re-wilding on a grand scale, but without mammoths such
efforts seem bound to fail, because mammoths and other elephants
are the ecological bankers of our world. Their grazing, defecation
and ploughing of snow in winter was vital to the entire ecosystem
of the mammoth steppe and allowed it to be vastly more productive
than it would otherwise have been.

Could we, should we, bring back the mammoth? The answer
to the 'could we' is 'not quite yet'. While scientists have made
advances in reconstructing the mammoth genome (as well as the
Neanderthal's and the thylacine's), they're still far from being able
to produce a living mammoth.[248] And the moral and ethical dimen-
sions of such science are daunting. We might give birth to a kind of
Frankenstein's monster, a genetic freak that could never live in the
real world. So why should we want to try? Simply because creatures
such as mammoths are vital elements in important ecosystems, and
it is only through restoring them that Earth's productivity and resilience can also be brought back to the level that would most benefit
our living planet, and thus ourselves. Attempting to re-establish
their role in ecosystems is thus similar to restoring health to our
farms and rangelands. It's also akin to helping a crashed economy
back onto its feet. The alternative is for humanity to remain eternal
'estate managers' of expansive, semi-wild regions whose contribution to the Gaian whole remain suboptimal.

If the future I've outlined is not merely fantastical, it has the
potential to herald a profound change in Gaia. From her birth
until now she has been a loosely coordinated entity lacking a

command-and-control system—a mere commonwealth of virtue—
and therefore unable to regulate herself precisely. But if the global
human superorganism survives and evolves, its surveillance systems
and initiatives to optimise ecosystem function raise the prospect of
an intelligent Earth—an Earth that would, through her global
human superorganism, be able to foresee malfunction, instability
or other danger, and to act with precision. If that is ever achieved,
the greatest transformation in the history of our planet would have
occurred, for Earth would then be able to act as if it were, as Francis
Bacon put it all those centuries ago, 'one entire, perfect living
creature'. Then the Gaia of the classical world would in fact exist.

James Lovelock believes that Gaia is already very old, frail, and
susceptible to human-caused upsets. But the Gaia that emerges
from this study is more akin to a newborn babe. All newborns
have new-formed brains, nervous systems and bodies, but these are
yet to be fully integrated so self-control and self-awareness remain
rudimentary. Infancy is the most dangerous period of life, and the
threats to our global civilisation that must be faced during this
century of decision will provide challenges enough, I think. But in
some future age, if our world is healed, our population stable and
a sustainable lifestyle established, the focus of our superorganism
will perhaps shift to the heavens. As the fortieth anniversary of
the Moon landing passed in 2009, it became evident that we have
retreated from the human exploration of space. No foot has trod
on a heavenly body since 1972, and there is no plan to return to the
Moon for at least a decade. This is perhaps appropriate. During
this critical period in the evolution of the human superorganism
all focus needs to be on Earth. But if we ever enter that long period
of stability that beckons from the far side of the crisis, we will
perhaps once again focus our energies on unlocking the mysteries of
the Universe.

Foremost among these mysteries is the question of whether or not there are other Gaias out there. The Italian physicist Enrico Fermi, in pondering the question, left us a paradox. It involves the simple question of why, despite the antiquity of the heavens and the vast number of stars and planets we know exist, have we not yet detected intelligent life?[249] There are 250 billion stars in the Milky Way alone, so surely some should have spawned Earth-like planets, and some of those should have developed life. Fermi assumed that it's a characteristic of life to colonise suitable habitats, its spread thereby making it more likely to be visible to us. If a civilisation used even the slow kind of interstellar travel almost within our grasp today, it would take only five million to fifty million years to colonise our galaxy. And that is just the blink of an eye in the fourteen-billion-year history of our Universe. Fermi's paradox would be resolved if infantile Gaias rarely, if ever, survive. If this is the explanation, then perhaps the Medea hypothesis is correct after all: intelligent global superorganisms may carry within themselves the seeds of their own destruction, and so begin to extinguish themselves at the moment of birth.

But there is another possibility. Perhaps Fermi's paradox tells us that we really are alone in the Universe, simply because we are the first global superorganism ever to exist. After all, it's taken all of time—from the Big Bang to the present—to make the stardust that forms all life, and to forge that stardust, through evolution by natural selection, into us and our living planet. If we really are the first intelligent superorganism, then perhaps we are destined to populate all of existence, and in so doing to fulfil Alfred Russel Wallace's vision of perfecting the human spirit in the vastness of the Universe. If we ever achieve that, then Gaia will have reached puberty, for she will then have become reproductive, nurturing the spark of life on one dead sphere after another. From our present vantage point we cannot know such things. But I am certain of one thing—if we do not strive to love one another, and to love our planet as much as we love ourselves, then no further human progress is possible here on Earth.

Acknowledgments

To David Flannery, my son, for alerting me to the marvellous Yan Fu, and for his many insights that have helped inform this book. To my daughter, Emma, and my spouse, Alexandra Szalay, who both see things far more clearly than I do, and who contributed in many ways to this book. To Michael Heyward, whose acute reading of several drafts saved me at least a decade in developing my ideas, and to Morgan Entrekin, who suggested the inclusion of an important new chapter. To Minik Rosing, for sharing Earth's oldest rocks with me, and to Nick Rowley, Peter Chapman, Vicki Flannery, Ed Shann, Frank Shann and Rob Purves, who read the manuscript and provided much correction.

Endnotes

1   Darwin, F. *The Life and Letters of Charles Darwin*, Vol. 8, The Echo Library, Teddington, Middlesex, 2007, p. 175.

2   Darwin, C. *The Formation of Vegetable Mould through the Action of Worms, with Observations on Their Habits*, John Murray, London, 1881.

3   ibid., p. 30.

4   Shermer, M. *In Darwin's Shadow: The Life and Science of Alfred Russel Wallace*, Oxford University Press, 2002, p. 118.

5   Bell, T. 1859, Anniversary Meeting, *Journal of the Proceedings of the Linnean Society, London*, 156: viii.

6   Darwin, C. 'On the Variation of Organic Beings in a State of Nature; on the Natural Means of Selection; on the Comparison of Domestic Races and True Species', extract from an unpublished *Work on Species*, read to the Linnean Society, London, 1 July 1858.

7   Wollaston, A. F. R. *Life of Alfred Newton: Late Professor of Comparative Anatomy, Cambridge University 1866–1907*, Dutton, New York, 1921, pp. 118–20.

8   Green, V. H. H. *A New History of Christianity*, Continuum, New York, 1996, p. 231.

9   Spencer, H. *The Principles of Biology (Vol. 1)*, University Press of the Pacific, Honolulu, 1864, pp. 444–74.

10   Flannery, D. 2009, 'Global Darwin: Ideas Blurred in Early Eastern Translations', *Nature*, 462: 984.

11   Darwin, C. 'On the Variation of Organic Beings in a State of Nature; on the Natural Means of Selection; on the Comparison of Domestic Races and True Species'.

12   Thatcher, M. Interview, *Woman's Own Magazine*, 31 October 1987, pp. 8–10.

13   Darwin, C. 'Recollections of the Development of My Mind and Character (1876–1881)', in Secord, J. A. (ed.) *Charles Darwin: Evolutionary Writings*, Oxford World Classics, 2009, p. 397.

14   Dawkins, R. *The Selfish Gene*, Oxford University Press, 1976, p. 21.

15   ibid., p. 2.

16   ibid., p. 139.

17   Semon, R. *The Mneme*, George Allen & Unwin, London, 1921;
      Semon, R. *Die Mnemischen Empfindungen*, William Engelmann,
      Leipzig, 1909.

18   Semon, R. *The Mneme*, p. 11.

19   ibid., p. 131.

20   ibid., p. 237.

21   ibid., p. 79.

22   Koestler, A. *The Case of the Midwife Toad*, Random House, New
      York, 1971.

23   Ward, P. *The Medea Hypothesis: Is Life on Earth Ultimately
      Self-destructive?*, Princeton University Press, 2009.

24   Wallace, A. R. 'On the Tendency of Varieties to Depart Indefinitely
      from the Original Type', Proceedings of the Linnean Society, London,
      1858.

25   Slotten, R. A. *The Heretic in Darwin's Court: The Life of Alfred Russel
      Wallace*, Columbia University Press, New York, 2004, p. 84.

26   Wallace, A. R. (5th edn) *Man's Place in the Universe: A Study of the
      Results of Scientific Research in Relation to the Unity or Plurality of
      Worlds*, George Bell & Sons, London & Bombay, 1905, pp. 243–61.

27   ibid.

28   ibid, pp. 258–9.

29   Gribbin, J. and Gribbin, M. *James Lovelock: In Search of Gaia*,
      Princeton University Press, 2009.

30   Lovelock, J. *Homage to Gaia: The Life of an Independent Scientist*,
      Oxford University Press, 2000, p. 253.

31   Lovelock, J. *Revenge of Gaia: Earth's Climate Crisis and the Fate of
      Humanity*, Westview Press, Boulder, 2007, p. 162.

32   Lovelock, J. 1972, 'Gaia as Seen through the Atmosphere', *Atmospheric
      Environment*, 6: 579–80.

33   Gribbin, J. and Gribbin, M. *James Lovelock*, p. 160.

34   Dawkins, R. *The Extended Phenotype: The Long Reach of the Gene*,
      Oxford University Press, 1982, pp. 234–6.

35   Staley, M. 2002, 'Darwinian Selection Leads to Gaia', *Journal of
      Theoretical Biology*, 218 (1): 35–46.

36   Golding, W. 'Gaia Lives, OK?', *Guardian*, 1976, cited in *A Moving
      Target*, Faber & Faber, London, 1982, p. 86.

37   Bacon, F. *Sylva Sylvarum*, William Lee, London, 1639.

38   Pell, Cardinal G. 2008, 'Global Warming and Pagan Emptiness: Cardinal Pell on the Latest Hysterical Subsitute for Religion', interview by M. Gilchrist, *The Catholic World Report*.

39   Cartigny, P., Harris, J. W. and Javoy, M. 1998, 'Eclogitic, Peridotitic and Metamorphic Diamonds and the Problem of Carbon Recycling—the Case of Orapa (Botswana)', in *Proceedings of the 7th International Kimberlite Conference*, Red Roof Design, Cape Town, 1: 117–24.

40   Bennett, V. 'Deep Time, Deep Earth: The Formation, Early History, and Large Scale Geochemical Evolution of the Earth', in *From Stars to Brains*, proceedings of a multi-disciplinary conference in honour of Paul Davies, Manning Clarke House, Canberra, Program & Abstracts, 2006, p. 29.

41   King James Bible, Genesis 3:19.

42   Rosing, M. T. et al., 2006, 'Consequences of the Rise of Continents—an Essay on the Geologic Photosynthesis', *Palaeogeography, Palaeoclimatology, Palaeoecology*, 232: 99–113.

43   ibid.

44   Nouvian, C. *The Deep: The Extraordinary Creatures of the Abyss*, University of Chicago Press, 2007.

45   Thomas, L. *The Medusa and the Snail: More Notes of a Biology Watcher*, Penguin, New York, 1995, p. 13.

46   Thomas, L. *The Lives of a Cell: Notes of a Biology Watcher*, Penguin, New York, 1978, p. 104.

47   Hamilton, W. D. and Lenton, T. M. 1998, 'Spora and Gaia: How Microbes Fly with Their Clouds', *Ethology, Ecology and Evolution*, 10: 1–16.

48   ibid.

49   Lenton, T. 'Hamilton and Gaia', in Ridley, M. (ed.), *The Narrow Roads of Gene Land*, Vol. 3, Oxford University Press, 2005, pp. 263–4.

50   Hamilton, W. D. 2000, 'My Intended Burial and Why', *Ethology, Ecology and Evolution*, 12: 111–12.

51   Wallace, A. R. 'On the Tendency of Varieties to Depart Indefinitely from the Original Type'.

52   Jantsch, E. *The Self-Organizing Universe: Scientific and Human Implications of the Emerging Paradigm of Evolution*, Pergamon, New York, 1980.

53   Darwin, C. *On the Various Contrivances by Which British and*

Foreign Orchids Are Fertilised by Insects, John Murray, London, 1862, pp. 197–8.

54 Thomas, E. M. *The Hidden Life of Deer*, HarperCollins, New York, 2009, pp. 170–1.

55 Wells, S. *The Journey of Man: A Genetic Odyssey*, Princeton University Press, 2002.

56 ibid, pp. 40–1.

57 Wells, S. *The Journey of Man*.

58 ibid.

59 ibid.

60 Martin, P. S. 'Prehistoric Overkill: The Global Model', in Martin, P. S. and Klein, R. G. (eds), *Quaternary Extinction: A Prehistoric Revolution*, University of Arizona Press, Tucson, 1984, pp. 354–403.

61 Yurtsev, B. A. 2000, 'The Pleistocene "Tundra-Steppe" and the Productivity Paradox: The Landscape Approach', *Quaternary Science Reviews*, 20: 165–74.

62 Guthrie, R. D. *Frozen Fauna of the Mammoth Steppe: The Story of Blue Babe*, University of Chicago Press, 1990.

63 Vasil'ev, S. A. 'Man and Mammoth in Pleistocene Siberia', in Cavarretta, G., Gioia, P., Mussi, M. and Palombo, M. R. (eds) *The World of Elephants: Proceedings of the First International Congress*, Consiglio Nazionale della Richerche, Rome, 2001, p. 2.

64 Martin, P. S. and Steadman, D. W. 'Prehistoric Extinctions on Islands and Continents', in MacPhee, R. (ed.), *Extinctions in Near Time: Causes, Contexts and Consequences*, Kluwer Academic, Plenum Publishers, New York, 1999, pp. 22–4.

65 Davis, O. K. and Shafer, D. S. 2005, 'Sporormiella Fungal Spores, a Palynological Means of Detecting Herbivore Density', *Palaeogeography, Palaeoclimatology, Palaeoecology*, 237: 40–50.

66 Flowers, S. E. Introduction and Commentary, *Ibn Fadlan's Travel-Report as It Concerns the Scandinavian Rûs*, Rûna-Raven Press, Smithville, Texas, 1998.

67 Morwood, M. and van Oosterzee, P. *A New Human: The Startling Discovery and Strange Story of the 'Hobbits' of Flores, Indonesia*, Smithsonian Books, Washington, 2007.

68 Hocknull, S. A. et al. 2009, 'Dragon's Paradise Lost: Palaeobiogeography, Evolution and Extinction of the Largest-ever Terrestrial Lizards (Varanidae).

69 Ashford, R. W. 2000, 'Parasites as Indicators of Human Biology and Evolution', *Journal of Medical Microbiology*, 49: 771–2.

70 Hansen, J. *Storms of My Grandchildren*, Bloomsbury, New York, 2009, p. 275.

71 Ostrom, E. *Governing the Commons*, Cambridge University Press, 1990.

72 Mitchell, T. *Journal of an Expedition into the Interior of Tropical Australia*, Longman, Brown, Green & Longmans, London, 1848.

73 Wilson, E. O. *Biophilia*, Harvard University Press, Cambridge, Massachusetts, 1984.

74 Fromm, E. *The Heart of Man: Its Genius for Good and Evil*, Harper & Row, San Francisco, 1965.

75 Marais, E. N. (1937), *The Soul of the White Ant*, Human& Rosseau, Cape Town, 2009, p. 151.

76 Hölldobler, B. and Wilson, E. O. *The Leafcutter Ants*, Norton, New York, 2010.

77 Hölldobler, B. and Wilson, E. O. *The Superorganism: The Beauty, Strangeness and Elegance of Insect Societies*, W. W. Norton, New York, 2009, p. 408.

78 ibid., p. 491.

79 ibid., p. 117.

80 ibid., p. 389.

81 ibid., p. 79.

82 Johnson, S. R. *Emergence: The Connected Lives of Ants, Brains, Cities and Software*, Scribner, New York, 2001.

83 Hamilton, W. D. 1964, 'The Genetical Evolution of Social Behaviour', *Journal of Theoretical Biology*, 7 (1): 1–52, p. 16.

84 Gagneux, P. et al. 1999, 'Mitochondrial Sequences Show Diverse Evolution Histories of African Hominoids', *Science*, 96 (9): 5077–82.

85 Smith, A. (1776), *An Inquiry into the Nature and Causes of the Wealth of Nations*, book 1, chapter 1, Oxford University Press, 1993.

86 Diamond, J. *Guns, Germs, and Steel*, W. W. Norton, London, 1997.

87 Smith, K. V. *King Bungaree*, Kangaroo Press, Sydney, 1992, p. 148.

88 Groves, C. P. 1999, 'The Advantages and Disadvantages of Being Domesticated', *Perspectives in Human Biology*, 491: 1–12.

89 Henneberg, M. 1988, 'Decrease of Human Skull Size in the Holocene', *Human Biology*, 60: 395–405.

90   Smith, A. *An Inquiry into the Nature and Causes of the Wealth of Nations*, book 5, chapter 1.

91   Butler, S. (1903), *The Way of All Flesh*, Penguin Classics, London, 1986, p. 59.

92   Callaway, E. 4 June 2009, 'Ancient Warfare: Fighting for the Greater Good', *New Scientist*.

93   Tench, W. *A Complete Account of the Settlement at Port Jackson*, London, 1793, in Flannery, T. (ed.), *Watkin Tench's 1788*, Text Publishing, Melbourne, 2009, pp. 246–7.

94   ibid., p. 117.

95   ibid., pp. 135–6.

96   FBI 2008, 'Crime Rate in the United States', www.fbi.gov/ucr/cius2008/data/table_01.html

97   Winter, J., Parker, B. and Habeck, M. (eds) *The Great War and the 20th Century*, Yale University Press, New Haven, 2000, p. 193.

98   Marschalck, P. *Bevölkerungsgeschichte Deutschlands im 19. und 20. Jahrhundert* Suhrkamp, Frankfurt, 1984, cited in 'World War II Casualties', Wikipedia, http://en.wikipedia.org/wiki/World_War_II_casualties

99   Hawkes, N. 2009, 'Conflict over War Deaths', Straight Statistics, www.straightstatistics.org/article/conflict-over-war-deaths

100   Plato, *The Republic*, translated by Lee, D., Penguin Classics (2nd edn), London, 1980.

101   Churchill, W. *Hansard*, London, 11 November 1947.

102   Plato, *The Republic*.

103   Diamond, J. *Guns, Germs, and Steel*.

104   Denham, T. P. et al. 2003, 'Origins of Agriculture at Kuk Swamp in the Highlands of New Guinea', *Science* 301: 189–93.

105   Flannery, T. 1991, 'Australia, Overpopulated or Last Frontier?', *Australian Natural History* 23: 769–75.

106   Smith, B. D. 2005, 'The Origin of Agriculture in the Americas', *Evolutionary Anthropology* 3: 174–84.

107   Flannery, T. *The Eternal Frontier: An Ecological History of North America and Its Peoples*, Text Publishing, Melbourne, 2001.

108   Bacon, F. *Novum Organum*, book 1, cxxix, translated by Spedding, J., 1858,

109   Levy, T. E. et al. 2002, 'Early Bronze Age metallurgy: A Newly

Discovered Copper Manufactory in Southern Jordan', *Antiquity* 76: 425–37.

110 Angela, A. *A Day in the Life of Ancient Rome*, Europa Editions, New York, 2009, pp. 80–6.

111 ibid., p. 337.

112 Moorhead, S. and Stuttard, D. *AD 410: The Year That Shook Rome*, British Museum Press, London, 2010.

113 Gribbin, J. R. *The Fellowship: The Story of a Revolution*, Allen Lane, London, 2005, p. 37.

114 Díaz del Castillo, B. *The Discovery and Conquest of Mexico, 1517–1521*, translated by Maudslay, A. P., Grove Press, New York, 1958.

115 Diamond, J. *Guns, Germs, and Steel*, p. 375.

116 Ruddiman, W. *Plows, Plagues and Petroleum: How Humans Took Control of Climate*, Princeton University Press, 2005, pp. 84–8.

117 Smyth, H. D. *Atomic Energy 1940–1945: A General Account of the Development of Methods of Using Atomic Energy for Military Purposes under the Auspices of the United States Government*, His Majesty's Stationery Office, London, 1945, p. 2.

118 Collins, P. 10 May 2005, 'Polar Eclipse. Hey, Remember When Climate Change Was a Swell Idea? Coconuts were in the offing', *Village Voice*.

119 9 February 1959, 'Canada: A-bombing for oil', *Time Magazine*.

120 ibid.

121 Fleming, J. R. 'On the Possibilities of Climate Control in 1962: Harry Wexler on Geoengineering and Ozone Destruction', American Geophysical Union, Fall Meeting 2007, abstract #GC52A-01.

122 Borisov, P. M. *Can Man Change the Climate?*, Progress Publishers, Moscow, 1973, p. 88.

123 ibid., p. 160.

124 Bhardwaj, R. et al. 2006, 'Neocortical Neurogenesis in Humans is Restricted to Development', *PNAS* 103 (33): 12564–8.

125 Koslow, T. *The Silent Deep: The Discovery, Ecology and Conservation of the Deep Sea*, UNSW Press, Sydney, 2007, pp. 148–9.

126 ibid., p. 140.

127 Carson, R. (1962), *Silent Spring*, Houghton Mifflin, New York, 2002, pp. 155–6.

128 Pope, J., Rosen, C. and Skurky-Thomas, M. 2010, 'Toxicity, Organochlorine Pesticides', *Emedicine*; Ballweg, M. L. *The Endometriosis*

Sourcebook, Macgraw Hill, Columbus, 1995, pp. 377–98.

129 Carson, R. *Silent Spring*, pp. 85–102.

130 ibid., p. 87.

131 ibid., p. 89.

132 ibid., p. 126.

133 ibid., pp. 7, 30.

134 ibid., pp. 118–19.

135 Pope, J. et al. 2010, 'Toxicity, Organochlorine Pesticides', *Emedicine*.

136 Carlsen, E., Giwercman, A., Keiding, N. and Skakkebæk, N. 1995, 'Declining Semen Quality and Increasing Incidence of Testicular Cancer: Is There a Common Cause?', *Environmental Health Perspectives* 103, Supplement 7, pp. 137–9.

137 Brown, R. M. 1947, 'The Toxicity of the "Arochlors"', *Chemist-Analyst* 36: 33.

138 Koslow, T. *The Silent Deep*, pp. 145–6.

139 ibid., p. 150.

140 National Environmental Trust, Physicians for Social Responsibility and Learning Disabilities Association of America, 2000, 'Polluting Our Future: Chemical Pollution in the US That Affects Child Development and Learning', p. 2, American Association on Intellectual and Developmental Disabilities, www.aaidd.org/ehi/media/polluting_report.pdf

141 Shapiro, J. *Mao's War against Nature*, Cambridge University Press, 2001.

142 ibid., p. 198.

143 ibid., p. 33.

144 Stockholm Convention on Persistent Organic Pollutants (POPs), http://chm.pops.int

145 Houde, M. et al. 2006, 'Biological Monitoring of Polyfluoroalkyl Substances: A Review', *Environmental Science & Technology* 40 (11): 3463–73.

146 Johns Hopkins University Bloomberg School of Public Health 2007, 'Polyfluoroalkyl Exposure Associated with Lower Birth Weight and Size', ScienceDaily, www.sciencedaily.com/releases/2007/08/070817115631.htm

147 Bagla, P. 25 February 2003, '"Mysterious Plague" Spurs India Vulture Die-off', *National Geographic*.

148 Roy, S. 15 April 2008, 'Vultures on Kolkata Skyline, Parsis Rejoice', Merinews, 2008, www.merinews.com/article/vultures-on-kolkata-skyline-parsis-rejoice/132527.shtml

149 Birdlife International, 22 May 2006, 'Second Blow for Asian Vultures', www.birdlife.org/news/news/2009/12/vultures.html

150 Koslow, T. *The Silent Deep*, p. 156.

151 American University 1997, 'TED Case Studies: Minamata Disaster', www1.american.edu/ted/minamata.htm

152 ibid.

153 Pacyna, E. G., Pacyna, J. M., Steenhuisen, F. and Wilson, S. 2006, 'Global Anthropogenic Emissions of Mercury to the Atmosphere', *Atmosphere Environment* 40 (22): 4048.

154 Cubby, B. 14 August 2008, 'Toxic Levels of Pollution Threaten River', *Sydney Morning Herald*.

155 Koslow, T. *The Silent Deep*, pp. 153–5.

156 ibid., pp. 154–5.

157 Pacyna E. G. et al. 'Global Anthropogenic Emissions of Mercury to the Atmosphere'.

158 Järup, L. et al. 1998, 'Health Effects of Cadmium Exposure—a Review of the Literature and a Risk Estimate', *Scandinavian Journal of Work, Environment and Health* 24: 11–51.

159 Nevin, R. 2007, 'Understanding international Crime Trends: The Legacy of Preschool Lead Exposure', *Environmental Research*, 104 (3): 315–36.

160 Brown, A. and Susser, E. 2008, 'Prenatal Nutritional Deficiency and Risk of Adult Schizophrenia', *Schizophrenia Bulletin* 34 (6): 1054–63.

161 Taylor, M. P. and Schniering, C. 2010, 'The Public Minimization of the Risks Associated with Environmental Lead Exposure and Elevated Blood Lead Levels in Children, Mount Isa, Queensland, Australia', *Archives of Environmental and Occupational Health* 65 (1):45–7.

162 Queensland Health 2007, 'Mount Isa Lead Report', www.health.qld.gov.au/news/media_releases/mtisa_leadreport.pdf

163 Marks, K. 12 September 2009, 'Living under a Cloud', *Sydney Morning Herald*, Good Weekend, p. 20.

164 Dayton, L. 27 February 2008, 'Mt Isa Heavy Metals "Still Toxic"', *Australian*.

165 Taylor, M. P. and Schniering, C. 'The Public Minimization of the

Risks Associated with Environmental Lead Exposure and Elevated Blood Lead Levels in Children, Mount Isa, Queensland, Australia'.

166 Xstrata, 22 June 2008, 'The Facts about Lead in Blood', *Sunday Mail*, p. 16 (advertisement).

167 Goldberg, E. D. 1986, 'TBT: An Environmental Dilemma', *Environment* 28: 17–44.

168 Nagy, B. et al. 1993, 'Role of Organic Matter in the Proterozoic Oklo Natural Fission Reactors, Gabon, Africa', *Geology* 21 (7): 655–8.

169 Bernstein, J. *Plutonium: A History of the World's Most Dangerous Element*, UNSW Press, Sydney, 2007.

170 Intergovernmental Panel on Climate Change 2007, 'Climate Change 2007: The Physical Science Basis', IPCC Fourth Assessment Report, www.ipcc.ch/ipccreports/ar4-wg1.htm

171 Richardson, K. et al. 2009, 'Synthesis Report from "Climate Change: Global Risks, Challenges & Decisions"', University of Copenhagen, www.climatecongress.ku.dk/pdf/synthesisreport.

172 ibid.

173 Hansen, J. *Storms of My Grandchildren*, pp. 83–4.

174 Malthus, T. R. (1798), *An Essay on the Principle of Population*, Oxford World's Classics, 2004, p. 61.

175 United Nations Department of Economic and Social Affairs, Population Division 2009, 'World Population Prospects: The 2008 Revision', http://esa.un.org/unpd/wpp2008/index.htm

176 Hull, T. H. 1990, 'Recent Trends in Sex Ratios at Birth in China', *Population and Development Review* 16: 63–83.

177 United Nations 2009, 'World Population Prospects: The 2008 Revision Population Database', http://esa,un.org/UNPP.

178 WWF 2008, 'Living Planet Report 2008', http://assets.panda.org/downloads/living_planet_report_2008.pdf

179 Benhabib, J. H. et al. 2010, 'Present-bias, Quasi-hyperbolic Discounting and Fixed Costs', *Games and Economic Behavior* 69: 205–23.

180 Milinski, M. et al. 2008, 'The Collective-risk Social Dilemma and the Prevention of Simulated Dangerous Climate Change', *Proceedings of the National Academy of Sciences of the United States of America* 105 (7): 2291–4.

181 Hobbes, T. *Leviathan, or the Matter, Forme and Power of a Commonwealth Ecclesiasticall and Civil*, Andrew Crooke, London, 1651.

182 Thaler, R. H. and Sunstein, C. R. *Nudge: Improving Decisions about Health, Wealth and Happiness*, Yale University Press, New Haven, 2008.

183 Kirby, K. N. et al. 1999, 'Heroin Addicts Have Higher Discount Rates for Delayed Rewards than Non-drug-using Controls', *Journal of Experimental Psychology: General* 128 (1): 78–87.

184 Daly, M. and Wilson, M. 2005, 'Carpe Diem: Adaptation and Devaluing the Future', *The Quarterly Review of Biology* 80 (1): 55–60.

185 Latham, R. C. and Matthews, W. (eds) (1661), *The Diary of Samuel Pepys*, HarperCollins, London, 1971.

186 Gribbin, J. and Gribbin, M. *James Lovelock*, p. 71.

187 Bernstein, P. L. *Against the Gods: The Remarkable Story of Risk*, John Wiley & Sons, New York, 1996.

188 Frank, R. H. *Passions within Reason: The Strategic Role of the Emotions*, W. W. Norton & Company, New York, 1988.

189 Frank, R. H., Gilovich, T. and Regan, D. T. 1993, 'Does Studying Economics Inhibit Cooperation?', *Journal of Economic Perspectives* 7 (2): 159–71, p. 170.

190 Ridley, M. *The Origins of Virtue*, Viking, London, 1996, p. 146.

191 Smith, A. *An Inquiry into the Nature and Causes of the Wealth of Nations*.

192 Lee, D. 2010, 'Beyond Simple Numbers: The Value of Environmental, Social and Governance Factors', *Alumni Association UQ Business School*, 2: 2–3.

193 Stern, N. *The Economics of Climate Change*, Cambridge University Press, 2007.

194 Generation Investment Management LLP, 2004, 'Generation Announces Plans for New Investment Management Firm', www.generationim.com/media/pdf-generation-final-launch-release-08-11-04.pdf

195 Balsam, S. *An Introduction to Executive Compensation*, Academic Press, San Diego, 2002, p. 135.

196 Levitt, S. D. and Dubner, S. J. *SuperFreakonomics: Global Cooling, Patriotic Prostitutes, and Why Suicide Bombers Should Buy Life Insurance*, William Morrow, New York, 2009, p. 171.

197 Cameron, J. and Blood, D. 'Catalysing Capital towards the Low-carbon Economy', *Thought Leadership Series No 3*, Copenhagen Climate Council, 2008.

198 ibid.

199 Monks, R. 2009, 'The Past, Present, and Future of Shareholder Activism: Bob Monks' Perspective', Speech, Harvard Law School, Cambridge, Massachusetts.

200 Principles for Responsible Investment 2009, 'PRI/UNEP FI Universal Owner Project: Addressing Externalities through Collaborative Shareholder Engagement: Request for Proposals', www.unpri.org/files/PRI_UNEP_FI_UO_Project_RFP.pdf

201 Monks, R. 2003, 'To Harvard with Love', Robert A. G. Monks, http://ragmonks.blogspot.com/2003/10/to-harvard-with-love.html

202 PRI Academic Network, http://academic.unpri.org

203 Friedman, T. *The World Is Flat: A Brief History of the Twenty-First Century*, Farrar, Straus & Giroux, New York, 2005.

204 Green, S. *Good Value*, Grove, New York, 2010.

205 http://worldbank.org

206 Windsor, J. 2006, 'Freedom in Africa Today', Freedom House, www.freedomhouse.org/uploads/special_report/36.pdf

207 International Bank for Reconstruction and Development/The World Bank 2008, '2005 International Comparison Program: Tables of Final Results', The World Bank, http://siteresources.worldbank.org/ICPINT/Resources/ICP_final-results.pdf

208 Green, S. *Good Value*.

209 Cameron, J. and Blood, D. 'Catalysing Capital towards the Low-carbon Economy'.

210 World Congress on Computers in Agriculture, 22–24 June 2009, program, Reno, Nevada.

211 Ion, A. 13 January 2009, 'Malaysia to Use Satellite Monitoring to Stop Illegal Logging', GreenPacks, www.greenpacks.org/2009/01/13/malaysia-to-use-satellite-monitoring-to-stop-illegal-logging

212 Carrington, D. 30 July, 2010, 'Satellite Sensors Help Cut Tree Felling in Amazon Rainforest', *Guardian Weekly* p. 3.

213 Fukuyama, F. *The End of History and the Last Man*, Free Press, New York, 1992.

214 Flannery, T. *The Weather Makers*, Text Publishing, Melbourne, 2005.

215 Milinski M. et al. 'The Collective-risk Social Dilemma and the Prevention of Simulated Dangerous Climate Change'.

216 Inman, M. 29 October 2009, 'The Climate Change Game', Nature Reports Climate Change, www.nature.com/climate/2009/0911/full/climate.2009.112.html

217 Safina, C. 2008, 'Regulators Are Pushing Bluefin Tuna to the Brink', Yale Environment 360 (Yale School of Forestry and Environmental Studies), http://e360.yale.edu/content/feature.msp?id=2096

218 Hardin, G. 1968, 'The Tragedy of the Commons', *Science* 162: 1243–8.

219 Ostrom, E. *Governing the Commons*.

220 Lovejoy, T., Flannery T. and Steiner, A. 27 October 2008, 'We Did It, We Can Undo It', *International Herald Tribune*.

221 Bacon, F. *Sylva Sylvarum*, p. 43.

222 Lewis, S. et al. 2009, 'Increasing Carbon Storage in Intact African Tropical Forest', *Nature* 457: 1003–6.

223 Wallace, A. R. *The Malay Archipelago: The Land of the Orang-utan and the Bird of Paradise*, MacMillan & Co., London, 1869, p. 35.

224 Lehmann, J. et al. 'Stability of Biochar in Soil', in Lehman, J. and Joseph, S. *Biochar for Enviromental Management*, Earthscan, London, 2009.

225 Van Zwieten, L. et al., 'Biochar and Emissions of Non-CO_2 Greenhouse Gases from Soil', in Lehman, J. and Joseph, S. *Biochar for Enviromental Management*.

226 Lehman, J. and Joseph, S. 'Stability of Biochar in Soil', in Lehman, J. and Joseph, S. *Biochar for Enviromental Management*.

227 Baker, J. M., Ochsner, T., Venterea, R. and Griffis, T. 2006, 'Tillage and Soil Carbon Sequestration—What Do We Really Know?', *Agriculture, Ecosystems and Environment* 118 (1–4): 1–5.

228 Kumar, R., Pandey, S. and Pandey, A. 2006, 'Plant Roots and Carbon Sequestration', *Current Science* 91 (7): 885–90.

229 Lal, R. 2007, 'Carbon Management in Agricultural Soils', *Mitigation and Adoption Strategies for Global Change* 12 (2): 303–22.

230 ibid.

231 Kumar, R. et al. 'Plant Roots and Carbon Sequestration'.

232 ibid.

233 Dong, X. 'Rangeland Soil Carbon Sequestration: The Contribution of Plant Roots', in North Dakota Agricultural Experiment Station 2007, '2007 CGREC Grass and Beef Research Review', North Dakota State University, www.ag.ndsu.edu/archive/streeter/2007report/carbon_seq/Carbon_Sequestration.htm

234 ibid.

235 Patton, B. D., Dong, X., Nyren, P. and Nyren A. 2007, 'Effects of Grazing Intensity, Precipitation, and Temperature on Forage

Production', *Rangeland Ecology and Management* 60 (6): 656–65.

236 Lal, R. 'Carbon Management in Agricultural Soils'.

237 Faragher, D. 2008, 'Kimberley Fire Management Team Takes Out Top Prize at WA Environment Awards', Government of Western Australia, www.mediastatements.wa.gov.au/Pages/Results.aspx?ItemID=130736

238 Burrows, N. D., Burbidge, A. A. and Fuller, P. J. 'Integrating Indigenous Knowledge of Wildland Fire and Western Technology to Conserve Biodiversity in an Australian Desert', in Foran, B. and Walker, B. (eds) *Science and Technology for Aboriginal Development*, CSIRO, Melbourne, and Centre for Appropriate Technology, Alice Springs, 1986.

239 Heckbert, S. et al. 2008, 'Land Management for Emissions Offsets on Indigenous Lands', CSIRO, www.csiro.au/resources/Indigenous-lands-emissions-offsets.html

240 ibid., p. 9.

241 Savory, A. *Holistic Resource Management*, Island Press, Washington, 1988.

242 Derner, J. D. and Schuman, G. E. 2007, 'Carbon Sequestration and Rangelands: A Synthesis of Land Management and Precipitation Effects', *Journal of Soil and Water Conservation* 62 (2): 77–85.

243 Bourke, T. 5 February 2010, address to forum on sustainable agriculture, Raheen, Melbourne.

244 Jowit, J. 9 April 2010. 'British Campaigner Urges UN to Accept "Ecocide" as International Crime', *Guardian*.

245 Leopold, A. *Round River*, Oxford University Press, 1972, p. 165.

246 Griffiths, J. *Wild: An Elemental Journey*, Hamish Hamilton, London, 2007.

247 Kizilova, A. 21 January 2007, 'Good Fence for Future Mammoth Steppes', Russia IC, www.russia-ic.com/education_science/science/breakthrough/357

248 Green, R. E. et al. 2008, 'A Complete Neandertal Mitochondrial Genome Sequence Determined by High-throughput Sequencing', *Cell* 134 (3): 416–26.
Miller et al., 2009, 'The mitochondrial genome sequence for the Tasmanian tiger (*Thylacinus cynocephalus*)', *Genome Research* 19: 213–20.

249 Davies, P. *The Eerie Silence: Are We Alone in the Universe?* Penguin, London, 2010, pp. 117–26.

Index

He just wanted a decent book to read ...

Not too much to ask, is it? It was in 1935 when Allen Lane, Managing Director of Bodley Head Publishers, stood on a platform at Exeter railway station looking for something good to read on his journey back to London. His choice was limited to popular magazines and poor-quality paperbacks – the same choice faced every day by the vast majority of readers, few of whom could afford hardbacks. Lane's disappointment and subsequent anger at the range of books generally available led him to found a company – and change the world.

'We believed in the existence in this country of a vast reading public for intelligent books at a low price, and staked everything on it'
Sir Allen Lane, 1902–1970, founder of Penguin Books

The quality paperback had arrived – and not just in bookshops. Lane was adamant that his Penguins should appear in chain stores and tobacconists, and should cost no more than a packet of cigarettes.

Reading habits (and cigarette prices) have changed since 1935, but Penguin still believes in publishing the best books for everybody to enjoy. We still believe that good design costs no more than bad design, and we still believe that quality books published passionately and responsibly make the world a better place.

So wherever you see the little bird – whether it's on a piece of prize-winning literary fiction or a celebrity autobiography, political tour de force or historical masterpiece, a serial-killer thriller, reference book, world classic or a piece of pure escapism – you can bet that it represents the very best that the genre has to offer.

Whatever you like to read – trust Penguin.